VALENCY:

CLASSICAL AND MODERN

By the same Author:

EXPERIMENTAL PHYSICAL CHEMISTRY

EXPERIMENTAL INORGANIC CHEMISTRY

VALENCY

CLASSICAL AND MODERN

BY

W. G. PALMER, Sc.D., D.Sc.

Fellow of St John's College, Cambridge

SECOND EDITION

CAMBRIDGE
AT THE UNIVERSITY PRESS
1963

CAMBRIDGE UNIVERSITY PRESS
Cambridge, New York, Melbourne, Madrid, Cape Town, Singapore, São Paulo, Delhi

Cambridge University Press
The Edinburgh Building, Cambridge CB2 8RU, UK

Published in the United States of America by Cambridge University Press, New York

www.cambridge.org
Information on this title: www.cambridge.org/9780521104968

This Edition © Cambridge University Press 1959

First edition 1944
Reprinted 1945, 1946, 1948
Second edition 1959
Reprinted 1963
This digitally printed version 2009

A catalogue record for this publication is available from the British Library

ISBN 978-0-521-05905-3 hardback
ISBN 978-0-521-10496-8 paperback

CONTENTS

PREFACE

It will be generally agreed that current opinions on the long-established conception of chemical valency, and on the ways in which it may be satisfied in chemical combination, originated in the attempts of Kossel, Lewis, and Langmuir (1916–1918) to derive a *modus operandi* from the contemporary ideas of atomic electronic structure. Since that time many manuals on the theory of valency have appeared, but they all expound principally, if not exclusively, the modern electronic views. To understand how a great and permanent theory first arose in the minds of its pioneers, how their successors contributed to its progress, or hindered it by misconceptions; to appraise justly how different the judgment of time and experimental application may be from the firm opinion of contemporaries, will always retain intense human, not less than scientific, interest. But in the subject of valency and chemical combination at least, there is the over-riding consideration that the rapid progress of modern theories would have been impossible, except upon the excellently laid foundations of pure chemistry. A volume attempting to exhibit the recent advances portrayed against the chemical background would therefore seem to be not inopportune at the present time, even though the modern theories are still far from complete, and many of the facts ascertained in classical chemistry await more than tentative explanations.

The subject of valency alone being as wide as the detailed chemistry of all the elements, not excluding carbon, it is essential to select for discussion in a work of limited size, and one intended for the use of students, only the main features and problems, and to confine the exemplifying compounds as far as possible to simple molecules. Although their quantitative powers cannot justly be ignored, I have made an effort to present electronic theories with the minimum of mathematical detail. At the same time some account of the spectroscopic basis of the present view of atomic structure has been regarded as necessary, in order to show how far the modern theory is independent of chemical reasoning. For I fear that two opposite kinds of false impression

may be taken from some recent expositions; either that the idea
of valency was coeval with the acceptance of the nuclear atom;
or, on the other hand, that modern physical theories take all their
cues from chemical facts, and are no more than an alternative
mode of restating those facts. While it would be narrow-minded
for physical theorists to be scrupulous to the point of ignoring
chemical reasoning, I have made it one object of the later
chapters of this book to demonstrate that they have indeed
given solutions to problems, which, first brought to light in pure
chemistry, by their very nature were insoluble by chemistry
unaided.

   After a short introductory chapter outlining the first tentative theories of chemical combination and the discovery of
valency as an empirical conception, the second and the first part
of the third chapter are concerned with the methods of determining chemical structure and valency; modern physical
methods as well as the classical chemical methods are discussed
and briefly explained. The second section of Chapter III illustrates the application of such methods to compounds of the
commoner elements treated according to the arrangement in
the periodic system. Up to this point all discussion follows
classical lines, though aided and supported by the results of
recently developed physical experimental methods. Only when
the problems and anomalies arising insistently from such a
treatment are thus clearly defined is the electronic theory introduced (in Chapter IV). This separation of the established chemical
problem from the current electronic solution is deliberate, and,
it is hoped, will not only set in a clear light the urgent need for
a comprehensive theory of chemical combination and valency,
but will permit the theory to display its power, without con-
cealing its present defects. For the sake of simplicity only the
'light' elements (hydrogen to fluorine) and their important compounds are discussed in Chapter IV, and the basic ideas of the
electronic theory are presented in a simplified form. The first
four chapters thus present a compact but elementary account
of classical and modern conceptions of valency, suitable for the
more general reader, and not overstraining the capacity of
candidates for University examinations of the standard of

Part I of the Natural Sciences Tripos at Cambridge. The concluding chapters are more advanced in content and treatment. Chapter v shows how the electronic theory is applied to the heavier elements and their principal compounds, while the last chapter is devoted to current developments and theories still in their early stages.

Much of any merit the book may be found to possess is due to the generous help I have gratefully received from many colleagues at Cambridge; notably from Mr A. J. Berry, Dr F. G. Mann, Dr G. B. B. M. Sutherland and Dr M. L. Tomlinson, all of whom have been ready with most useful suggestions for improvements. I count myself fortunate also in having had encouragement and sage advice from my former teacher, Dr W. H. Mills, now President of the Chemical Society. I feel bound to add that any controversial opinions are advanced on my own responsibility, and must not be regarded as sponsored or condoned by any of these friends. All such friendly support from colleagues would hardly have sufficed to achieve publication, had I not also benefited from the long-suffering collaboration of the staff of the Cambridge University Press, to whose forbearance under the often excruciating difficulties of the times I shall ever be indebted.

I must plead guilty, not only to enlarging an already voluminous terminology, but to including in this expansion the objectionable word 'co-ionic' (Chapter iv). In mitigation I offer the excuse that great masters of chemistry were formerly content to use such terms as 'monovalent' and 'pentavalent'; in our own time acknowledged leaders have not despised 'electrovalent' and 'electronegativity', or even 'ionicness'. To avoid over-burdening the text with references to the literature, I have, when a recent communication supplies adequate bibliography, given as a rule only the recent reference. I hope to be pardoned, if, owing to the difficulties under which the book was prepared, I have failed to notice any important relevant work of the present day.

W. G. P.

April 1944

PREFACE TO THE SECOND EDITION

During the decade preceding the War of 1939–45 the progress of chemistry had been vastly accelerated by the impetus gained on the one hand from the newly explored metrical methods of X-ray and electron diffraction, and on the other from the brilliant expansion of theoretical insight into what had remained for so long the mystery of chemical union. The first edition of this book in 1944 appeared after an inevitable period of slackening during the war years, and at a time when it was not easy to be sure that all the most recent advances had been noted.

The first task therefore in preparing a new edition has been to incorporate the more precise molecular measurements secured since the end of the war, not only by improvements in applying the older methods, but by new techniques, such as neutron-diffraction, micro-wave spectroscopy, and nuclear magnetic resonance. Re-validated thus, and by the correction of some surviving errors, the first four chapters remain 'a compact but elementary account of classical and modern conceptions of valency, suitable for the general reader'. Chapter v also retains broadly its original form, but now includes a short account of 'artificial' and transuranic elements.

In Chapter vi substantial changes have proved imperative: the accounts of the metallic carbonyls, of acetylene, of the electro-affinity (electronegativity) of combined atoms, and of the energies of single links have all been entirely rewritten: the section on molecular orbits has been modernized, and amplified by new matter on electron deficient systems, such as the boron hydrides. A summary of the recent treatment of the *cyclo*alkanes has been introduced, and it is hoped that the now detailed knowledge of 'hydrogen bonding' is reflected in the greatly expanded section on this very wide-spread type of molecular interaction.

I must not fail to offer my thanks to certain of my colleagues, notably Dr A. G. Sharpe, Dr P. Maitland and Dr A. G. Maddock, who by reading and commenting upon drafts of new or amended sections have so greatly assisted in the attempt to ensure a correct and acceptable presentation.   W. G. P.

LIST OF ABBREVIATED TITLES

Acta Cryst. Acta Crystallographica.
Amer. Abstr. Abstracts of the American Chemical Society.
Angew. Chem. Angewandte Chemie.
Annalen. Liebig's Annalen der Chemie.
Ann. Chem. Annalen der Chemie und Pharmacie.
Ann. Physik. Annalen der Physik.
Ann. Reports. Annual Reports of the Chemical Society.
Atti Accad. Lincei. Atti della Accademia nazionale dei Lincei: Rendiconti della Classe di Scienze fisiche, matematiche e naturali.
Ber. Berichte der deutschen chemischen Gesellschaft.
Bur. Stand. J. Res. Bureau of Standards Journal of Research (to 1935).
Chem. Ind. Chemistry and Industry.
Chem. Rev. Chemical Reviews (American Chemical Society).
Compt. rend. Comptes rendus hebdomadaires des séances de l'Académie des Sciences.
I.C.T. International Critical Tables.
J. Amer. Chem. Soc. Journal of the American Chemical Society.
J. Chem. Physics. The Journal of Chemical Physics.
J. Chem. Soc. Journal of the Chemical Society.
J. Ind. Eng. Chem. Journal of Industrial and Engineering Chemistry.
J. Physical Chem. The Journal of Physical Chemistry.
J. Res. Nat. Bur. Stand. Journal of Research of the National Bureau of Standards (from 1935).
Naturwiss. Die Naturwissenschaften.
Phil. Mag. Philosophical Magazine.
Phil. Trans. Philosophical Transactions of the Royal Society of London.
Phys. Rev. Physical Review.
Pogg. Ann. Poggendorff's Annalen (Annalen der Physik).
Proc. Roy. Soc. Proceedings of the Royal Society.
Proc. Roy. Soc. Edin. Proceedings of the Royal Society of Edinburgh.
Quart. Rev. Chemical Society Quarterly Reviews.
Rec. trav. chem. Recueil des travaux chimiques des Pays-Bas et de la Belgique.
Rev. Mod. Phys. Reviews of Modern Physics.
Trans. Faraday Soc. Transactions of the Faraday Society.
Verh. dtsch. phys. Ges. Verhandlungen der deutschen physikalischen Gesellschaft.
Z. anorg. Chem. Zeitschrift für anorganische und allgemeine Chemie.
Z. Elektroch. Zeitschrift für Elektrochemie.
Z. Krist. Zeitschrift für Kristallographie.
Z. Physik. Zeitschrift für Physik.
Z. physikal. Chem. Zeitschrift für physikalische Chemie, Stöchiometrie, und Verwandtschaftslehre.

Chapter I

THE ORIGINS

1. Dalton

While science is pursuing a steady onward movement, it is convenient
from time to time to cast a glance back on the route already traversed,
and especially to consider new conceptions which aim at discovering the
general meaning of the stock of facts accumulated from day to day in
our laboratories.

MENDELEEFF, Faraday Lecture of the Chemical Society,
J. Chem. Soc. 55, 634 (1889).

Had Mendeleeff been lecturing on the theory of valency he
would doubtless have chosen a stronger word than 'convenient',
for unless the reasons for the long delay in its discovery are
properly assessed and appreciated, an essential perspective is
wanting, and the dominance attained by the theory in later
chemistry will hide its subtlety, and unexpectedness. It is
tantalizing to speculate whether this central doctrine of modern
chemistry would have emerged much earlier, if Dalton had been
more sensitive to the chemical aspects of the atomic theory,
Avogadro's principle been more generally accepted at the date
of its first proposal, and atomic weights fixed by Berzelius in
place of Gerhardt and Cannizzaro. An affirmative answer is
improbable, for the hesitation of the earlier generations of
chemists to accept the hypotheses that to-day are seen in so
clear a light was mainly due, not to any obtuseness or prejudice,
but rather to the lack of a secular maturation of both thought
and experiment, which, inherited in the end by their successors,
enabled the latter to progress at a hitherto unparalleled rate.

That the discoverer of the law of multiple proportions should
have failed to realize some of the implications of the law in
regard to the subdivision and limitation of combining power is
traceable in part to Dalton's preoccupation with physical
matters, in particular with the serious problems of gaseous
diffusion arising from the Newtonian picture of gaseous constitution still held in Dalton's time. A gas was conceived of as
consisting of a sparse distribution of heavy particles kept

separated by a relatively voluminous but tenuous envelope of 'caloric' or heat. The elastic properties of a gas were ascribed to the compressibility of the caloric envelope. On withdrawal of heat the system contracted, liquefaction and finally solidification ensuing. When Dalton and his contemporaries wrote of the size of a gaseous particle they meant the diameter including the caloric envelope. To a world regarding a kinetic theory of matter as almost axiomatic this entirely static picture seems fantastic, but when the single idea of molecular motion was unknown it gave a workable, and it may be the only possible picture. It is too seldom realized that those who laid the experimental foundations of our knowledge of gases—Priestley, Gay-Lussac, Dalton, Avogadro—all thought of a gas in some such way.

Dalton's early interest in meteorology (see footnote)* had brought him into contact with the properties of those gaseous mixtures that form the atmosphere; and he had, as is well known, sufficiently studied the solubility of gaseous mixtures to enunciate a law (1802). He saw that the fact, unknown to Newton, that the atmosphere was a mixture of gases, would demand, on the static Newtonian picture, a stratification, the denser particles accumulating in immediate contact with the earth. Such an effect would of course have results quite distinct from the well-known gradual change of density with height and pressure. In 1803 Dalton discussed such problems in a memoir entitled *Spontaneous Intercourse of Different Elastic Fluids in confined Circumstances* (in modern terms—Gaseous diffusion). He tentatively advanced the novel theory that spontaneous mixing might be due to the inequality of the 'sizes' of particles of gases differing in the chemical sense, suggesting that the smaller particles might penetrate between the larger, and, as he said, 'an intestine motion' be set up. It was obvious that if the relative weights w of different particles could be found, the (relative) number N per unit volume of gas was immediately calculable as $N = d/w$ (d = specific gravity), and then the relative diameter as $1/\sqrt[3]{N}$.

In Dalton's mind the atomic theory had at this time become the basis of his mode of thought, and he now associated with the

* *Meteorological Observations and Essays*, London, 1793.

general theory certain rules of chemical combination, and in so doing took an epoch-making step in the history of chemistry. His most important rules may be thus summarized: (a) if two elements A and B yield only one compound, its composition is assumed to be AB, (b) if more than one compound is known, there will be A_2B and AB_2 in addition to AB, and so on for still more complex unions. In fact he took a purely statistical view of the probability of formulae, regarding A_2B and AB_2 as equally likely. Here was certainly no inkling of the principle of valency, but it is not to be forgotten that a number of Dalton's contemporaries with livelier chemical intuition, notably Gay-Lussac, Wollaston and Davy, expressed scepticism of these arbitrary rules (see e.g. Wollaston, *Phil. Trans.* **104**, 1 (1814)). Using existing analyses mainly of compounds he regarded as of the type AB (e.g. water HO, sulphuretted hydrogen HS, carbonic oxide CO, ammonia NH, phosphoretted hydrogen PH) he prepared (1803) his first table of 'atomic weights' (H = 1), including of course what we should now term molecular weights. Later, Dalton himself experimented on the composition of nitrous and nitric oxides, and of methane and ethylene, in order to secure an acceptable experimental basis for his rules of combination, which however he himself probably regarded as self-evident. The law of multiple proportions resulted.

Having prepared a table of relative weights, Dalton proceeded to his main object, the calculation of 'atomic sizes', and produced the table given in full below.

Table 1

Compound (modern formulae)	Specific gravity (air = 1)	Diameter	Compound (modern formulae)	Specific gravity (air = 1)	Diameter
SiF_4	—	1·15	Cl_2	—	0·981
HCl	—	1·12	NO	1·102	0·980
CO	—	0·94	SO_2	2·265	0·95
CO_2	1·500	1·00	N_2O	1·610	0·947
H_2S	1·106	1·00	NH_3	0·580	0·909
PH_3	—	1·00	C_2H_4	—	0·81
H_2	0·077	1·00	O_2	1·127	0·794
CH_4	—	1·00	N_2	0·966	0·747

Atomic weights (1810)

Hydrogen	1	Carbon	5·4
Oxygen	7	Sulphur	13
Nitrogen	5	Phosphorus	9

During the course of this work Dalton noted but could not explain the facts that although the 'atom' of water must be heavier than that of oxygen, and that of nitric oxide nearly twice as heavy as the 'atom' of oxygen, yet steam was much lighter than oxygen and nitric oxide of about the same density. By assigning the formulae H_2O, NO and O_2 in 1811 Avogadro provided a complete explanation, which however failed to convince Dalton (see below).

In the table are recorded Dalton's atomic weights of 1810, those specific gravities that he left on record, and his final calculation of diameters, given in a lecture in 1810. For convenience and brevity the gases are listed under modern formulae. It will be seen that five different gases appear to have identical sizes, and the general variation from equality is much less than the variation of either the 'atomic weight' or the density. Yet Dalton, perhaps a little biased by his theory of diffusion, refused to see anything suggestive in this near-equality. He in fact regarded the values of the table as finally refuting the claims of Avogadro (1811) and others to interpret Gay-Lussac's law of gaseous combination (1809) as proving the equal 'sizes' of all gaseous particles, or as is now said 'equal volumes contain equal numbers of particles'. It is an interesting speculation that had Dalton been of a rather more sanguine temperament he might have given to the world not only the atomic theory, but also that generalization, now associated with the name of Avogadro, which yields in importance only to the atomic theory. At least he would have withheld that bitter opposition, which, combined with his great prestige, more than any other single influence militated for fifty years against the general acceptance of Avogadro's principle.

The following quotations illuminate Dalton's tendency to adopt a deductive attitude, and also the movement of chemical thought in his day. (Italics have been inserted.)

The combination of gases in equal or multiple volumes is naturally connected with this subject. Cases of this kind, or at least approximations to them, frequently occur; but no principle has been suggested to account for the phenomena; *till this is done I think we ought to investigate the facts with great care: and not suffer ourselves to be led to adopt analyses till some reason can be discovered for them.*

New System, Vol. II, 1827.

When he [Dalton] did see it [Gay-Lussac's account of his law of gaseous combination, in 1809] he set himself to show that Gay-Lussac's opinions were ill-founded. But the later researches of chemists have left no doubts about their accuracy; and if Mr Dalton still withholds his assent, he is, I believe, the only living chemist to do so.

System of Chemistry, edition of 1825 by Thomas Thomson. In the edition of 1807 Dalton encouraged Thomson to publish a preliminary summary of the atomic theory.

2. Berzelius

Dalton could have had no more wholehearted supporter of his ideas than Berzelius, who patiently devoted the early years of a long life of chemical research to discovering and perfecting methods of analysis, ultimately demonstrating beyond all doubt the validity of the law of multiple proportions. Berzelius, in his approach to the problems of chemical composition, made full use of the results of his pupil Mitscherlich in the new field of isomorphism (1819), and also of the law of atomic heats announced by Dulong and Petit in 1819. His analyses soon compelled him to modify Dalton's original rules of chemical union; thus on finding the oxygen in two oxides of iron to be in the proportion 2:3, he changed Dalton's formulae FeO and FeO_2 to FeO_2 and FeO_3; later an application of the principle of isomorphism forced a further modification to the modern formulae FeO and Fe_2O_3.

During the first decade of the last century, it had been established that by electrolysis the results of chemical combination can often be disunited. As examples we may cite the electrolysis of water (without addition of acid or alkali) by Nicholson and Carlisle in 1800; the decomposition of salts into acid and base by Berzelius and Hisinger in 1803; and, most famous of all, the isolation of metals from the caustic alkalis in 1807 by Davy using the electrolysis of fused alkali, electrolysis of saturated solutions having failed to produce the metal. The notion that 'chemical action is reversed electrolysis', to be so insisted upon in later years by the late Professor Armstrong, was already present in many minds, but it was especially Berzelius who devised a working electrochemical theory of chemical union (1812–1819). Elementary atoms were supposed to possess

2

(unspecified) quantities of both positive and negative electricity, but usually not in equal amounts; in electronegative, i.e. anode-seeking elements, the negative predominated, and conversely for cathode-seeking or electropositive elements. The supporters of this theory never made clear how it could be compatible with the electroneutrality of matter in its normal state. Both kinds of electricity seemed to be needed in each atom. First, in order to take account of the formation of more than one compound between atoms A and B; it appeared necessary, for example, to explain the two chlorides taken to be CuCl and $CuCl_2$ by assuming that copper could adopt two electropositive states, in which the balance between the two kinds of electricity was different. Secondly, to explain how, for example, sulphur could be negative in CaS and positive in SO_2. Chemical combination was simply regarded as the partial mutual saturation of the residual charges on the combining elements, which were necessarily of electropositive and electronegative character respectively. The idea of a *saturation* (Berzelius used the term 'neutralization') rather than of mere electrostatic attraction was demanded even to ensure a constant composition for a given chemical individual; and in this we may see, with some risk of committing an anachronism, the first germ of a theory of valency.

The *partial* nature of the saturation was essential to the theory. In bases (basic oxides), such as sodium, calcium and copper oxides, the original positive electrification of the metal predominated; in acids (acid anhydrides) the negative character of oxygen had the upper hand. Hence the possibility of salt formation; and, if the electrification were not then finally neutralized, double-salt formation could follow. All chemical compounds were regarded as at least formally dissociable into two oppositely charged parts, an attitude summarized in the description 'dualistic'. The theory afforded a classification of acids and bases—bases such as the oxides of sodium and calcium which, owing to the high net positive charge on the metal, still retained much positive charge were classed as 'strong'; acids were 'strong' if the negative charge was marked. Since the negative charge on the acids was ascribed to oxygen, acids with a high proportion of oxygen were strong. It is noteworthy that, with the modern meaning of strength, such a classification is still in the main valid—as examples, we may take sulphurous and sulphuric, nitrous and nitric, hypochlorous and perchloric. The most stable compounds were naturally those formed from two 'strong' components: stability and charge were interrelated.

This first electrochemical theory of combination is typical of a class of hypotheses, that by their very elasticity excite scepticism; they seem to explain everything but predict nothing. Whether the second, modern, electrochemical theory is also in this class may be a matter for argument, but at least it may be urged that it rests on a far more quantitative basis than the first. It was upon a quantitative question that the theory of Berzelius came finally to grief. In 1834 Faraday announced his laws of electrolysis, which show that the same amount of electricity is needed to release an atom of copper from its sulphate as to liberate two atoms of hydrogen from a molecule of water; atomic charge and stability—or as we now might say, affinity—are not related. Other contemporaneous developments damaging to the theory were the discovery, associated with the name of Dumas, that the electronegative atom chlorine could replace the electropositive hydrogen in organic compounds without fundamental change of properties, and the growing acceptance of the polyatomicity of elementary molecules such as those of the common elements (p. 12). Had the theory of substitution been seen against a more general background than that of the compounds of carbon, less weight would have attached to it, for PH_3 and PCl_3 are certainly bodies of very distinct properties. Although the theory had to be abandoned, the ideas on which it was based became a permanent acquisition to chemistry. It raised for the first time the *nature* of chemical union, and rightly ascribed it to electrical action; it postulated rightly, although in an entirely speculative way, the presence of both kinds of electricity in atoms, and it assumed that mere electrostatic attraction between the combining units is insufficient to explain the facts of combination.

The difficulty about elementary molecules, such as H_2 and Cl_2, was of course that no polar mechanism could be assumed to act between identical atoms. In this matter Berzelius allowed himself to take a grossly illogical position; he accepted Avogadro's doctrine for compounds, but rejected it for elements. We must, however, not allow longstanding familiarity with the 'doubled' formula to blind ourselves to the fact that a real explanation of the union of like atoms was first forthcoming from the work of Heitler and London in 1927, which was based on the modern subject of quantum mechanics. A point of similarity in the scientific attitude of Dalton and Berzelius may be detected, in that both men refused to accept apparently well-founded experimental facts because these facts appeared to controvert preconceived theories. Before real progress could be achieved in understanding the nature of chemical union, the most difficult step had to be taken—the disentanglement of the *existence* of what we should now call a bond, from the *strength* of the bond. In the series HCl, HBr, HI it is assumed that in all cases a single bond unites the elements, but, as we now well recognize, the bond *strength* decreases regularly in the series, in a non-integral manner. The distinction here being drawn, in effect between valency and affinity, is entirely unexpected and at first sight unplausible. We should rather have expected the very reactive chlorine to unite with more atoms of hydrogen than the less reactive iodine. That the difficulty was long felt is shown by the fact that A. W. Hofmann, in his *Einleitung in die moderne Chemie*, edition of 1877, after a carefully worded warning (on p. 183) against confusing 'intensity of chemical avidity' with 'quantivalence', proceeds to use words and phrases such as 'Bindekraft' and 'Kraft des Sauerstoffatoms' (on p. 269), and thus seems himself, at least to a literal reader, to confuse 'bond force' with bond. The first to draw public attention to the need for distinction was Frankland, whose ideas form part of the subject-matter of the following section.

3.   Frankland and Kekulé

In the *Phil. Trans.* of 1852, at p. 440, at the close of a paper describing the preparation and properties of novel organometallic compounds including (in modern formulae):

CH_3HgI	$(C_2H_5)_2SnO$	$(C_2H_5)_2Sn$
$(C_2H_5)_2SnI_2$	$(C_2H_5)_2SnS$	$(C_2H_5)_4Sn$
	$(C_2H_5)_3SbO$	$(C_2H_5)_3Sb$

Frankland writes as follows: 'When the formulae of inorganic chemical compounds are considered, even a superficial observer is struck with the general symmetry of their construction; the compounds of nitrogen, phosphorus, antimony and arsenic especially exhibit the tendency of these elements to form compounds containing 3 or 5 equivalents of other elements, and it is in these proportions that their affinities are best satisfied; thus in the ternal group we have NO_3, NH_3, NS_3, PO_3, PH_3, PCl_3, SbO_3, SbH_3, $SbCl_3$, AsO_3, AsH_3, $AsCl_3$, etc., and in the five-atom group NO_5, NH_4O, NH_4I, PO_5, PH_4I, etc. Without offering any hypothesis regarding the cause of this symmetrical grouping of atoms, it is sufficiently evident from the examples just given that such a tendency or law prevails, and that, no matter what the character of the uniting atoms may be, the combining power of the attracting element, if I may be allowed the term, is always satisfied by the same number of these atoms.'

(The formulae are expressed in Gmelin's equivalent notation $O = 8$, $H = 1$.)

Referring to the series of antimony compounds, which may be expressed in modern formulae as follows:

$$Sb(C_2H_5)_3 \qquad Sb(C_2H_5)_4OH \qquad Sb(C_2H_5)_3(OH)_2$$

Frankland says:* 'Stibethine furnishes us, therefore, with a remarkable example of the operation of the law of symmetrical combination above alluded to, and shows that the formation of a five-atom group from one containing three atoms, can be effected by the assimilation of two atoms, either of the same, or of opposite electrochemical character: this remarkable circumstance suggests the following question—is this behaviour common also to the corresponding compounds of arsenic, phosphorus and nitrogen; and can the position of each of the five atoms with which these elements respectively combine be occupied indifferently by an electronegative or an electropositive element?' We discern in the first of these excerpts the first clear conception of valency; and in the second, some final hesitation to contravene the surviving dualistic hypothesis. It was particularly propitious that the idea of valency should have

* *Loc. cit.* p. 422.

been first promulgated in connection with *metallic* compounds, for although the increasing recognition of the value of vapour and gaseous densities in fixing molecular weights and thence molecular formulae had been very generally applied to *volatile* compounds, metallic compounds, of which few are volatile, were still largely regarded as amenable to Dalton's arbitrary rules; in particular it had been customary to ascribe metallic oxides to the binary type MO (NaO, CaO, CuO, SbO, FeO, etc.), any appearance of specific valency being thus hopelessly obscured.

On reading Frankland's communication impartially a doubt must arise whether he realized the far-reaching importance of his conceptions, and his aloofness from controversies and participation in later developments does not lessen that doubt. In this he differed from Kekulé, who in his *Lehrbuch der org. Chemie*, p. 115 (1861), gave a clear discussion of valency, with special reference to the valency of carbon, without however in any way acknowledging a debt to Frankland. Whatever the true circumstances may be, Kekulé certainly soon became the dominating figure in the rapid expansion of the new ideas. With characteristic energy (and no little dogmatism) Kekulé proceeded to reorganize the already large bulk of knowledge in organic chemistry. In this direction he could successfully assume that in nearly all organic compounds hydrogen, oxygen, nitrogen and carbon had the constant valencies 1, 2, 3 and 4 respectively. He explained chain formation in the following words: '...the carbon atoms are themselves linked together, whereby a part of the affinity of the one carbon atom is necessarily tied by an equally large part of the affinity of the other',* and showed that $2n + 2$ is the saturation capacity of C_n. The possibility that the valency of carbon was less than 4 in unsaturated compounds was held by most of Kekulé's contemporaries (e.g. Wurtz, Kolbe, Erlenmeyer, Gerhardt, Williamson, and Couper).† The following extract from Williamson's *Chemistry for Students*, edition 1868, p. 126, indicates his cautious attitude:

'When carbon is burnt to form carbon monoxide each mole-cule of the oxide contains one quadrivalent atom of carbon

---

* *Ann. Chem.* **106**, 154 (1858).
† For Couper's views see *Compt. rend.* **46**, 1157 (1858).

united with one divalent atom of oxygen; the equivalence is therefore different from usual; we might say that carbon is divalent or oxygen quadrivalent. One element undergoes a change of valency on combustion to carbon dioxide. When olefiant gas is completely burnt,

$$C_2H_4 + 3O_2 = 2CO_2 + 2H_2O,$$

four univalent atoms of hydrogen are replaced by four divalent oxygen atoms, and the apparent equivalence of carbon is doubled. In the partial combustion,

$$C_2H_4 + 2O_2 = 2CO + 2H_2O,$$

four hydrogen atoms are replaced by two divalent oxygen atoms, and the carbon apparently continues divalent.' Williamson does not discuss these questions further, but he evidently means to imply that he considers carbon has the same valency in CO as in C_2H_4, and that this valency is probably two.

Kekulé, having proposed, with general assent, a single bond between identical carbon atoms to explain carbon chains, saw no reason why multiple bonds should not also occur between carbon atoms as they undoubtedly did between carbon and oxygen or carbon and nitrogen; and by this assumption he could maintain the constant quadrivalence of carbon, otherwise prevailing in organic chemistry. Kekulé's brilliant application of his idea of multiple bonds to the problem of benzene* convinced most of his contemporaries of its correctness. At this time and until recently the two bonds in ethylene were regarded as equivalent and identical in nature, whatever this was, but this assumption has now been questioned in relation to modern theories (p. 193). Also, in relation to modern views on bond-lengths (p. 64) it is interesting to note that Kekulé referred to multiple linkages as 'more compact' forms of union than the single bond.† When he applied his principle of constant valency to elements other than carbon, Kekulé met with obvious difficulties. He was compelled to write phosphoric pentachloride as $PCl_3.Cl_2$ and ammonium chloride as $NH_3.HCl$, calling these for the first time *molecular compounds*.‡ The dissociability of these

---

* *Ann. Chem.* 137, 129–96 (1866).   † *Ann. Chem.* 106, 156 (1858).
‡ *Compt. rend.* 58, 510 (1864).

bodies appeared to support such views. The structure $P\!\!\begin{array}{c}\diagup\text{OCl}\\\!-\text{Cl}\\\diagdown\text{Cl}\end{array}$
for phosphorus oxychloride had less experimental evidence in
its favour. Modern theories in every way support Kekulé's
long championship of the constant valency of elements in what
is now recognized as the first series in the periodic classification,
which includes carbon, hydrogen, oxygen and nitrogen (IV,
pp. 112–16).

4. Gerhardt, Clausius and Cannizzaro

Neither Gerhardt nor Cannizzaro can be said to have made
direct contributions either to the theory of valency or of
chemical union in general, but by their steady support of the
principle of gaseous volumes they finally succeeded in esta-
blishing molecular and atomic weights on a sound and un-
ambiguous basis; and this is an achievement necessarily preceding any successful theories of either valency or chemical
combination. The slow but continuous progress of the hypothesis
now connected with the name of Avogadro may be seen in the
following chronological notes:

1809 Gay-Lussac published his law of combining volumes,
and interpreted it to mean that equal volumes of gases contain
equal numbers of Daltonian atoms (Gay-Lussac would have
said that 'the sizes of the particles are equal', see p. 2).

1811 Avogadro re-interpreted Gay-Lussac's law, assuming
that the particles of the elementary gases were diatomic, and
in this year gave the following formulae:

$$H_2, O_2, N_2 \text{ and } Cl_2; \quad HCl, NH_3, N_2O \text{ and } NO;$$
$$CO_2, SO_2 \text{ and } SO_3;$$

in 1814 he added

$$H_2S \text{ and } CS_2; \quad CH_4, C_2H_4 \text{ and } COCl_2.$$

1814 Ampère stated the gas-volume hypothesis independently.

1827 Dumas accepted the hypothesis and furthered its
application by the invention of his method of determining
vapour densities.

1834 Prout regarded the hypothesis as proved.

1853 Gerhardt, in his *Traité de Chimie Organique*, says: 'All molecules in the state of gas should be made to occupy the same volume.' Gerhardt, of course, accepted the Newton-Dalton conception of gaseous constitution (p. 2). Translated into our terms the maxim means that molecular weights should be so chosen that the gram-molecular weights of all gases occupy the same volume (of course under like physical conditions). It is remarkable that Gerhardt attributes the hypothesis to Ampère. Unfortunately for the future of the gas-volume hypothesis Gerhardt was over-zealous, and regarded the molecules of *all* elementary bodies as diatomic, in particular those of mercury and phosphorus. Confusion was therefore only transferred from one class of elements to another.

Although Bernoulli in 1738 had adumbrated in a rather obscure way a kinetic theory of gases it was not until 1857 that the modern kinetic theory of matter was founded by Clausius, who, in his use of the gas-volume hypothesis, followed Dalton's lead in bringing chemistry to the support of physics.[*] It may easily be shown that the pressure exerted on its boundaries by an assemblage of moving particles is $p = \frac{1}{3}nmc^2$, where n is the average number of particles per unit volume, each of mass m, and c is the average speed. It is known from experiment that the pressure of a gas held at constant volume rises in proportion to $(t + 273)° = T°$, where t is the Centigrade temperature. Hence for a given mass nm of any *one* gas, T is proportional to c^2. For two dissimilar gases to be in stable physical equilibrium, both p and t must be the same; for this case $n_1 m_1 c_1^2 = n_2 m_2 c_2^2$. If we accept the gas-volume hypothesis $n_1 = n_2$, and therefore equality of temperature is associated with the equality $m_1 c_1^2 = m_2 c_2^2$, i.e. equal average kinetic energies. Once this stage is reached the way is open for a full development of the kinetic theory.

In 1858, Cannizzaro, in his now famous *Outline of a Course of Chemical Philosophy*, gave the first practical definition of atomic weight as 'the least weight of an element found in the gram-

[*] O. E. Meyer, *Die kinetische Theorie der Gasen*, English transl. 1899, p. 64; and Clausius, *Pogg. Ann.* **100**, 370 (1857); transl. in *Phil. Mag.* (4), **14**, 108 (1857).

molecular weight of its compounds'. In this statement he subordinated the atom to the molecule, the exact antithesis of Dalton's attitude. The very rapid way in which the leaven of Cannizzaro's ideas spread in chemistry is shown by the history of the periodic classification. In 1864 Newlands signally failed to convince his contemporaries of the correctness of the periodic principle, and his failure was due almost entirely to the faulty atomic weights he employed. Only five years later, in 1869, Mendeleeff achieved his brilliant success.

5. Mendeleeff

No advance contributed as much to the consolidation and development of the doctrine of valency as the establishment of the periodic system of classification. In this, it is valency which is essentially the periodic property, and this characteristic of atoms was seen for the first time as a natural feature rather than as merely a convenient method of classifying compounds. Mendeleeff deprecated the prominence given in the past to the doctrine of *constant* valency, and, rejecting in principle Kekulé's idea of molecular compounds (p. 11), boldly asserted that a given atom might exhibit different valencies towards the test elements hydrogen, chlorine and oxygen. He was also the first to notice that the sum of the separate valencies towards hydrogen and oxygen is eight, and emphasized that carbon had the special property of exhibiting equal valencies to all of the three test elements, hydrogen, chlorine and oxygen (II, p. 29). An account of the principal features of the periodic system will be found in the following chapter, and in Chapter v.

References. The following works are recommended:
Roscoe and Harden, *A New View of the Origin of Dalton's Atomic Theory*, London (1896).
Richard Anschütz, *A. W. Kekulé*, Berlin, 2 vols. (1929).
E. von Meyer, *A History of Chemistry*, transl. McGowan, Macmillan (1906).

Chapter II

DEVELOPMENT OF GENERAL THEORY TO THE RISE OF THE ELECTRONIC THEORY

In this chapter and the next an attempt will be made to present impartially the main content of the subject of valency and chemical union before the impact of the electronic theory, whose influence may be dated from Kossel's memoir of 1916. It is not possible, however, to assign a limiting date to the period which will be covered, since not the least of the benefits deriving from the new theories is the stimulus that has been received to a further and more searching examination of all kinds of valency problems, to many of which no solution has yet been given. In order to avoid confusion and false perspective no explicit references will usually be made in this chapter to electronic solutions, but cross-references *in italics* will indicate where in the succeeding chapters such solutions may be found. We shall now find it convenient, in showing the extension of ideas which arose in circumstances indicated in Chapter I, to abandon a chronological outlook, in favour of a summary treatment of developments which may have actually occupied many years.

Before entering upon this account, something must be said about nomenclature. Most of the pioneers of the theory of valency used the term 'atomicity' in the sense of valency; a few wrote of 'quantivalence' or 'equivalence'. The credit of perceiving the confusion that arose from the continued employment of such terms in more than one sense must be given to Hofmann who, in his text-book of 1865, advises that 'atomicity' be retained explicitly for the composition of elementary molecules, and first introduced the term 'Werthigkeit', translated as 'valency'. In recent times a confusion has again arisen between the words 'valence' and 'valency'; one reads of 'a valence' meaning a bond, of 'the valency' meaning combining power, while 'valency' is used indiscriminately in the strict numerical sense and in the more general sense of 'aptness for combination'. It is proposed in what follows to use 'valency' only in a numerical sense, and 'bond' in place of 'valence'. No compact symbolic way of expressing valency seems yet to have been devised. In what is to follow the expression $(X), XY = n$ should be read 'the valency of X in its compound XY is n'; when there is no need to mention a particular compound, the simpler expression $(X) = n$ will be used.

The experimental determination of valency

Units of valency, and test elements

Valency being a numerical quantity cannot be measured except in relation to a defined unit. Since the time of Kekulé it has been an almost universal custom to symbolize a valency unit by a single stroke or bond between the atoms, and structural formulae or *bond-diagrams* drawn on such a basis have long been familiar in organic chemistry. The term *bond-diagram* implies that emphasis is laid on the proper assignment of valency units, while a structural formula need only show as a minimum requirement, the actual juxtaposition of the atoms in the molecule. The valency n is of course equal to the number of bonds issuing in the bond-diagram from the atom concerned. Until quite recent times it was impossible to ascertain by *direct* experiment the number of bonds issuing from an atom in combination, but the modern treatment of *bond-lengths* allows some information to be directly gained in simple cases (III, p. 44). The classical chemical method aimed at bringing the element, say A, into union with one or more atoms of an element of *indivisible* combining power, say X, and so forming the compound AX_n. We may then be sure that n bonds issue from A, and state $(A), AX_n = n$. Two vitally important conditions must in this method be satisfied: (1) the compound must be known to contain only one atom of A, (2) it must be certain that the valency of X is 1. Hence the molecular as well as the empirical formula of AX_n must be known; and therefore a determination of valency by this method is most certain if AX_n is volatile. It will be immediately plain that were we to take a compound, say A_2X_n, the bond-diagram is not self-evident, and until a decision can be made about it, no definite value of the valency of A can be found. Thus, if we assume for the moment that hydrogen and fluorine are truly univalent, $(S), H_2S = 2$, and $(S), SF_6 = 6$, but $(S), S_2Cl_2$ will seem to depend on how the bond-diagram is drawn: in $\begin{smallmatrix} S-Cl \\ | \\ S-Cl \end{smallmatrix}$ we see bivalency, in $S{=}S{<}^{Cl}_{Cl}$ we confer valencies of 4 and 2 on the sulphur atoms. It would be reasonable to be guided in choosing a bond-diagram for S_2Cl_2 by the established bivalency in H_2S, but the chemical properties of sulphur chloride tend to favour the second formula (p. 83).

In deciding on the use of test elements of unit valency we meet with very serious difficulties, if we rely solely upon classical chemistry, as it is the aim of this chapter to do. It is one thing to assume arbitrarily, as the early exponents of valency did, that hydrogen is univalent; it is quite another to decide authoritatively whether in any circumstances its combining power can undergo division. The classical arguments may be shortly rehearsed as follows:

(1) *Compounds of hydrogen with one other atomic species* X *contain only one atom of* X *when they contain one atom of hydrogen.*

If we make a list of such compounds whose molecules can be *proved* to contain only one atom of hydrogen we find only HCl, HBr, HI and HN_3 (b.p. 39°). Compounds such as LiH, NaH and KH are of course non-volatile, and their molecular weights are indirectly deduced, even if they are to be regarded as known. We have hardly a clear demonstration here of the univalence of hydrogen, even though the accepted structure of HN_3 is of course written to show univalent hydrogen (III, p. 79).

(2) *The existence of the series of typical hydrides* HCl, H_2O, H_3N, *and* H_4C is difficult to understand unless $(H) = 1$, for if hydrogen were, for example, bivalent, the valencies of the other elements become multiples of 2, which seems unlikely.

(3) *In electrolysis no element or group has a less charge than hydrogen, although many have equal charges.*

The application of Faraday's laws of electrolysis provides an alternative method of determining valency, when the products of electrolysis are simple and elementary. Faraday showed conclusively that in the electrolytic liberation of elements quantities of electricity are concerned which are multiples of a single

18   CLASSICAL THEORY

unit, numerically equal to that associated with the release of
one gram-atom of hydrogen. In no known example of electrolysis does the electricity required to release a gram-atom of
hydrogen rise above the minimum of one unit, although it is
now known that the unit may be positive, as in the great majority
of electrolyses involving hydrogen, or negative, as in the
decomposition of LiH dissolved in fused $LiCl_2$.

The doubts that have from time to time been raised about the
univalency of hydrogen in all circumstances indicate that these
arguments cannot be taken as finally conclusive. Thus de Forcrand* could seriously propose bivalency as the normal condition
of hydrogen. The consequent doubling of previously accepted
valencies would then allow, by providing alternative bond-
diagrams, a constant valency to be maintained for most elements.
However, the fantastic nature of the diagrams proposed by
de Forcrand could hardly commend them to chemists; as examples we may quote

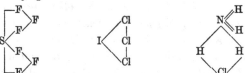

In more recent times the stable existence of the ion HF_2^- has been
established, and appears to demand a valency greater than one
for either hydrogen or fluorine. Modern theories would require
effectively bivalent hydrogen in unionized NH_4OH (which may
perhaps be regarded as a reason for denying the existence of this
compound). The apparently abnormal volatility of certain
o-hydroxy-aromatic compounds (o-nitrophenols, o-hydroxy-
aldehydes) has given rise to the suggestion that the hydroxy group
is linked to the other substituent group through hydrogen, which
thus becomes effectively bivalent. A similar explanation has been
put forward to explain the unexpected volatility of many enol
forms of tautomeric compounds (acetoacetic ester, acetyl-
acetone). In the opposite direction the unexpectedly low vapour
pressures of most alcohols have been attributed to association,

* *Compt. rend.* **140**, 764 (1905).

the linking atom being (bivalent) hydrogen. Finally, during the past few years examples have come to light during crystal analysis by means of X-ray diffraction in which two oxygen atoms appear to be so close that a bonding must be assumed, and the linking atom is considered to be hydrogen (the position of which is undetectable by X-ray methods, III, p. 38). It is difficult to see what alternative explanation can meet these cases, which will be discussed in more detail in a later chapter (*p. 222*), where it will be suggested that hydrogen may remain essentially univalent even in these examples.

Since the time of Berzelius the *mutual* character of chemical interaction has been recognized as unique among the interactions of matter (I, p. 6). In mechanical systems we find an active portion of matter, e.g. a compressed spring, exerting a force upon a second portion. The magnitude of the force depends in no way on the second portion, which will, however, by its movement give a measure of the force. In a similar unilateral way charged bodies attract insulators, the latter being passive partners in the action. The highly active fluorine has, however, no chemical attraction for helium or neon, and combines in no greater proportion with hydrogen than does iodine. We might hope to find some analogy in the electrical forces between two bodies both charged where there is a real mutual action, but attraction is here dependent on the opposed signs of the charges, and we are faced, like Berzelius, with the example of the very stable molecules of the elements (I, p. 7). If it were conceivable that the force of chemical attraction were exactly proportional to atomic weight, then we might trace an analogy with gravitation, but this is of course the very opposite of the truth, $H_2(M = 2)$ being for example far more stable than $I_2(M = 254)$. It is to be regarded as a major triumph for the modern electronic theory that it has at last provided an explanation of the nature of this elusive mutual effect. Without adopting any explanation at the moment for chemical interaction we may still incorporate its mutual character into a useful rule of valency—if a plane is imagined to be drawn between any two atoms in chemical union the valencies of the portions on each side of the plane are equal.

Thus on the basis $(H) = 1$, if we separate one hydrogen from the hydrocarbon $C_nH_{(2n+2)}$ we prove at once that the (alkyl) radicals $C_nH_{(2n+1)}$ are all univalent. An obvious extension proves that such groups as NO_2, NH_2, OH and CO_2H are also univalent. Realizing that hydrogen, owing to the paucity of stable hydrides, is not an entirely suitable standard of valency for direct use, we may apply the principle of mutual action to the hydrides of the very reactive halogens, and at once obtain (Cl), $HCl = 1$, wherein Cl may be replaced at will by Br or I. It would seem clear from this that in the requisite compound AX_n, X may be a halogen and the way is open for determining the valency of practically all the elements, as all except the inert elements form compounds with the halogens.

It was early recognized that the principle of substitution in organic chemistry, exemplified by the series CH_4, CH_3Cl, CH_2Cl_2, $CHCl_3$ and CCl_4, strongly supported the equi-valence of hydrogen and halogen. We may also state (C), $CH_4 = (C)$, $CCl_4 = 4$. But when we regard the set PH_3, PF_5 and SH_2, SF_6 the principle of substitution would indicate other conclusions, if we adhere to constant valency for each element, for the fluorides, unlike PCl_5, are the most stable known compounds of phosphorus and sulphur; they cannot in any sense be considered, as in Kekulé's suggestion (p. 11), as compounds of the type $PCl_3 . Cl_2$. It is known that at room temperature the density of hydrogen fluoride corresponds approximately to the formula H_3F_3, but in this, HF molecules, which are highly polar, might be held merely by dipole attractions, without necessitating a valency of F greater than 1. The existence of the ion HF_2^- in solution (p. 88) raises the question of the valency of F and H in more acute form. It will be shown later (*p. 222*) that there are reasons for thinking that $(F) = 1$ even in this ion.

The existence of a great variety of halogen oxides, and of interhalogen compounds of types higher than binary, raises difficult questions about the true valency of the halogens. In addition to the binary interhalogen compounds (e.g. ICl, ClF, BrF) the following compounds appear to be definite:

ClF_3 (b.p. 12°)	BrF_3 (b.p. 127°)	ICl_3, solid (m.p. 101°)
	BrF_5 (b.p. 40°)	IF_5 (b.p. 97°)
		IF_7 (b.p. 5–6°, m.p. 4·5°)

In addition to the above compounds the following compound ions are of established composition:

$$I_3^- \text{ and } Br_3^-, \quad IBr_2^-, ICl_2^-, ClBr_2^-, ICl_4^- \text{ and } IFCl_3^-.$$

Most of these ions are rather unstable, but the explanation suggested for H_3F_3—a linkage of dipoles—cannot be advanced for I_3^- or Br_3^-, and is unlikely for the other cases. The oxides Cl_2O_7 and I_2O_5 may be mentioned; the heptoxide appears to be actually the least unstable of the oxides of chlorine, while I_2O_5 is stable up to 200°. In the case of the oxides it is possible to construct bond-diagrams showing univalent halogen and bivalent oxygen. Thus we may follow Wurtz (1862) in proposing the structure I—O—O—O—O—O—I, but little chemical experience is necessary to reject such a diagram for a substance stable to 200°. Obviously such devices will not avail to retain univalency in the interhalogen compounds.

It will not have escaped the reader's notice that fluorine appears to have an exceptional power of exciting high apparent valencies; PF_5, SF_6, are very stable and show no tendency to thermal dissociation. IF_5 and OsF_8 are stable to 400°. This might suggest that the minimum valency of 1 must be assigned to fluorine, in order that the valency exerted by the other elements should not be excessive. Univalency may be assumed also in the oxide F_2O. It must, however, not be forgotten that were higher valency assigned to the halogen, bond-diagrams could be constructed which would not necessarily demand high valencies from the other elements in these fluorides (see for example de Forcrand's formulae, p. 18 and cf. Wurtz's formula for I_2O_5 above). A decision on these questions must therefore depend on a settlement of molecular structure (III, p. 47). The complicated situation of halogen valency may be summarized thus: in compounds of carbon, and in those of many elements of lower valency, halogens replace hydrogen atom by atom; we may instance LiH, LiCl;CaH_2, $CaCl_2$, and remember that the valency of the metals is independently proved by the electrochemical method (p. 17). On the other hand, it cannot possibly be asserted that the halogens are invariably univalent.

The almost universal formation of oxides by the elements

3   P V

would favour its employment as a test element for valency. Unfortunately its bivalency leads on the one hand to the existence of numerous classes of oxide, of distinctive chemical properties; and on the other to the possibility for a given oxide of alternative bond-diagrams. Mendeleeff, who was inclined to give more prominence to oxygen than to chlorine as a test element, laid down the empirical rule that only salt-forming oxides were to be considered. A salt is formed from a basic oxide by treatment with acid, say hydrochloric, and the only other product of the interaction is water. The general equation

$$\text{basic oxide} + \text{acid} = \text{salt} + \text{water,}$$

as for example

$$Fe_2O_3 + 6HCl = 2FeCl_3 + 3H_2O,$$

is commonly made the basis for chemical definitions of any two of the chemical types concerned. The reaction ensures that the chloride formed *corresponds* to the basic oxide, for if it does not water will not be the only other product. The use of basic oxides may be thus reduced to the use of chlorides, and the assumption of the univalency of chlorine. But it is admittedly arbitrary to write in the above equation $2FeCl_3$ in place of Fe_2Cl_6, which latter might well be held to correspond more exactly to Fe_2O_3. For the 'molecule' Fe_2Cl_6 we should probably construct the diagram $(Cl)_3{\equiv}Fe{-}Fe{\equiv}(Cl)_3,$

and hence for the oxide

and we are led to regard iron as at least quadrivalent. There is some evidence that ferric chloride, like aluminium chloride, is at least dimeric. It is, on the other hand, certain that the ion taking part in electrolysis of simple ferric salts is Fe^{+++} and not $(Fe_2)^{6+}$ (in contrast to the case of the mercurous ion, which is $(Hg_2)^{++}$ with like certainty (III, p. 58)). It will be clear from this one example that the use of basic oxides for the determination of metallic valency is not without ambiguities. Acidic oxides have very generally been regarded as the final products of dehydration of existent or hypothetical oxy-acids. The limiting forms of oxy-acids—the true ortho-acids—are expressed as $X(OH)_n$, where X is a non-metallic element. Only when n does not exceed 3, as in $B(OH)_3$, does the true ortho-acid easily survive the tendency to spontaneous dehydration, $P(OH)_5$ becoming $PO(OH)_3$, the actual 'ortho' phosphoric acid, $C(OH)_4$ becoming $CO(OH)_2$, etc. Esters of true ortho-acids are of course well known, as for example the ortho-carbonic esters. This accident of spontaneous dehydration does not, however, destroy the correspondence between the acidic oxide (acid anhydride) and the acid $X(OH)_n$. The use of acidic oxides is thus tantamount to the use of OH as a valency unit. It might be conjectured that in oxides of form R_2O and RO assignments of valencies 1 and 2 respectively might confidently be made. But the examples N_2O, NO and CO indicate the dangers of so simple a procedure, since the valencies of nitrogen and carbon are well established as 3 (or 5) and 4 respectively, not only by NH_3, CH_4, CCl_4 but by an enormous number of organic compounds. Further difficulties encountered in deciding on the valency exhibited in oxides will be discussed later, in the sections devoted to the periodic classification.

Before proceeding further, another (still largely unsolved) question of the general theory of valency must be mentioned. On general grounds we might expect a principle that those compounds would be most stable in which an element was exerting its combining power to the fullest advantage, and to discover a lessened stability when the apparent combining power is either less or greater than the principal valency. Such would be a plausible interpretation of Frankland's words (I, p. 9): '...and it is in these proportions [i.e. those indicated by the valencies] that their affinities are best satisfied'. If we take 'saturated compound' to mean one in which only single bonds occur, and 'unsaturated' to connote the presence of multiple bonds, the facts of organic chemistry are mainly in harmony with this view. The paraffin hydrocarbons probably represent the most stable group of organic compounds; in the olefines and their derivatives the *apparent* combining power of carbon is less than the quadrivalency, and marked chemical reactivity is found in this class.

Table 2. The periodic system

0	IA	IIA	IIIA	IVA	VA	VIA	VIIA	VIII	VIII	VIII	IB	IIB	IIIB	IVB	VB	VIB	VIIB	0
							A GROUPS							B GROUPS				
He 2																		
Ne 10	Li 3	Be 4	B 5											C 6	N 7	O 8	F 9	Ne 10
Ar 18	Na 11	Mg 12	Al 13											Si 14	P 15	S 16	Cl 17	Ar 18
Kr 36	K 19	Ca 20	Sc 21	Ti 22	V 23	Cr 24	Mn 25	Fe 26	Co 27	Ni 28	Cu 29	Zn 30	Ga 31	Ge 32	As 33	Se 34	Br 35	Kr 36
Xe 54	Rb 37	Sr 38	Y 39	Zr 40	Nb 41	Mo 42	Tc 43	Ru 44	Rh 45	Pd 46	Ag 47	Cd 48	In 49	Sn 50	Sb 51	Te 52	I 53	Xe 54
	Cs 55	Ba 56	La 57	[Ce 58]*														
			Rare earth metals 59–71															
Rn 86				Hf 72	Ta 73	W 74	Re 75	Os 76	Ir 77	Pt 78	Au 79	Hg 80	Tl 81	Pb 82	Bi 83	Po 84	At 85	Rn 86
	Fr 87	Ra 88	Ac 89	Th 90	Pa 91	U 92												

* Ce is 'transitional' to the rare-earth elements (pp. 72, 165).

The group $C{=}O$ which is capable of expansion to $\overset{\displaystyle |}{C}{-}\overset{\displaystyle |}{O}$ is the cause of reactivity in very many types of organic compounds; radicals such as CH_3 and CH_2 have only a transitory existence. The only exception to the principle among carbon compounds is carbon monoxide. When, however, we turn to the compounds of other elements the principle seems often inapplicable. The principal valency of nitrogen may be taken as 3, yet N_2O_3 is the only oxide of nitrogen the existence of which could still be doubted; in the chemical sense N_2O is probably the most stable oxide of nitrogen, i.e. the least reactive. When sodium reacts with oxygen at ordinary pressure, the chief product is the oxide Na_2O_2, and not Na_2O. With potassium under like conditions we do not obtain K_2O but an oxide of the formula KO_2 (cf. III, p. 50). The hydride BH_3 has never been isolated, and must be presumed to be too unstable for existence, but B_2H_6 and many other hydrides of boron are known in which the *apparent* combining power of boron is greater than 3 (cf. VI, p. 233). Chlorides of sulphur, S_2Cl_2, SCl_2 and SCl_4, are known, but only the common S_2Cl_2 is reasonably stable.

Valency and the periodic system of classification

Since Mendeleeff first promulgated the periodic system in 1869 many alternative modes of setting out the classification have been devised and advocated, in order to bring to light further relationships. But, short of the clarification of the system brought about by modern knowledge of atomic structure, which will be considered in a later chapter, it does not appear that anything essential is to be gained by modifying the classical mode of Mendeleeff, except by inserting elements discovered since his time. We shall also adhere for the present to a general ordering by atomic *weight*. The system is now thought to contain 92 positions for elements excluding the 'transuranic' metals (p. 165), of which all are certainly known, and of these some 70 possess metallic properties. Only in the case of three pairs of elements does the ordering by atomic weight break down, viz. Ar 39·9, K 39·1; Co 58·9, Ni 58·7; Te 127·6, I 126·9. The recurrent or periodic property is essentially *valency*. From a general survey of chemical and physical properties, among which the property of valency was not specially emphasized, we should probably often associate in 'local' groups elements placed in separate groups in the periodic system. As examples we may take the groups C, P and S; F and O; Cr, Mn, Fe (Co and Ni); 'the noble metals' Ag, Pt and Au; Cu, Hg; and the so-called 'diagonal relations' of Li to Mg, Be to Al, B to Si are well known. At some points what might appear on chemical grounds to be a subsidiary valency is given first place: (Cu) = 1, (Tl) = 3, (Au) = 1 and (Mn) = 7.

Periodicity is now recognized as more complex than a simple law of 'octaves'. The system consists of *short series*, uniformly containing eight elements, separated by *long series*, the length of which increases as atomic weight increases. The *short series* exhibit the most characteristic features of the system. They are all terminated by an inert element. Passing backwards, we find a halogen always contiguous to the inert element, followed in turn by three elements of rapidly decreasing electronegative character (non-metals in earlier series). Finally come three metallic elements (excepting boron in the first series). The *long periods* all begin, like the short periods, with three metals, the first always an alkali metal, so that all the inert elements have on the side of lower atomic weight (or atomic number) a halogen and on the other an alkali metal. The electronegative character terminating all the short series fails to appear in the long series, and these consist always of elements of predominantly metallic nature. The most interesting features of the *long series* may be illustrated from the first of such series, which extends inclusively from K to Ni:

(1) From Ti to Ni inclusive the majority of the compounds of the metals are *coloured*, as are the metallic kations.

(2) Elements near the middle of the series, V, Cr and Mn, exhibit *numerous types of compounds*, and the valency actually exerted depends mainly on whether the chemical environment is oxidizing or reducing.

(3) The cations of Cr, Mn, Fe, Co and Ni are *paramagnetic* (p. 158). The group of 'rare-earth' metals occurs in the middle of a very long period. Many of these show coloured compounds, and

all the cations are paramagnetic; in addition they exhibit a constant tervalency. These features may be contrasted with those of the short series, wherein, excepting those of the first element (Cu, Au), coloured compounds are rarely encountered, valency changes steadily and diamagnetism is the rule.*

Looking now at the vertical association, into *groups*, we find as before mentioned that valency is the controlling factor in the grouping. After the elements in the first two (short) series, all groups contain alternating members of long and short series. The chemical differences between the sub-groups thus formed are so marked that the sub-groups are commonly considered separately, and labelled A and B families respectively. Thus Ca, Sr, Ba and Ra fall into the A family of Group II, while Zn, Cd and Hg are assembled in the B family. Members of the B families are all located in short series.

One of the most significant differences between the A and B families is that in the former the valency is usually constant throughout, while in the B families valency shows a marked tendency to decrease with increasing atomic weight. For example, we may notice in the fourth group, that the heaviest member of the A sub-group, thorium, is still strongly quadrivalent, while in lead, the heaviest member of the corresponding B family, quadrivalency is vestigial. As a rule the metallic elements in the A families exhibit very simple and very similar crystallographic features, while much greater complexity is met with in the B-family metals. This distinction is of great importance for the physical behaviour of metals, and in metallographic classification. It is commonly said that in the A families there is a steady increase of electropositiveness with increasing atomic weight. This assertion is untrue in general if electropositiveness is gauged by the formation of ions *in water*.† The energy necessary to produce one gram-cation of a metal from the (gaseous) metal *in vacuo* (i.e. the ionization potential) and the corresponding energy to produce one gram-cation in water from the

* See for details, Chapter v, p. 158 et seq.

† See, however, p. 220 where there is a section on the electro-affinity of combined atoms.

crystalline metal are shown in Table 3. (ΔH = heat of formation, ΔF = free energy of formation.)

Table 3. *Energies of formation of ions (in electron-volts)*

	Li	Na	K	Rb	Cs
M$^+$ gas	5·36	5·11	4·32	4·16	3·87
*M$^+$ aq. $\begin{cases}\Delta H \\ \Delta F\end{cases}$	2·90 2·96	2·51 2·72	2·63 2·93	2·66 2·93	2·71 —
	Mg	Ca	Sr	Ba	
M^{++} gas	14·96	11·82	10·98	9·95	
*M^{++} aq. $\begin{cases}\Delta H \\ \Delta F\end{cases}$	4·80 5·30	5·65 5·96	5·67 —	5·57 —	
	F	Cl	Br	I	
*X$^-$ aq. $\begin{cases}\Delta H \\ \Delta F\end{cases}$	3·4 —	1·72 1·36	1·24 1·07	0·58 0·53	

* Values for aqueous ions are computed on the arbitrary basis that the heat and free energy of the reaction $\frac{1}{2}H_2 \to H^+$ aq. are zero (*I.C.T.* **5**, 176, and **7**, 224).

For the halogens is shown the energy necessary for the reaction X^- aq. $\to \frac{1}{2}X_2$ (gas). In this (B) family there is a rapid increase of electropositiveness even for the hydrated ions.

In Mendeleeff's classification the Group VIII contained, unlike all other groups, three sets of triads, each composed of chemically very similar metals (which belong crystallographically to the A families). As each triad seemed to form a region of coalescence between a long and short period the elements in Group VIII were termed 'transition elements'. The ambiguity in the position of the lightest element, hydrogen, gave rise to lively dispute, and in the course of time opinion swayed between placing it as the first halogen, or as the first alkali metal; its assumed univalency precludes other positions. It is hardly worth while at the present time to rehearse the various arguments used to support the one or the other position (see *p. 110*).

Looking broadly at the periodic scheme shown on p. 29 we may discern a regular association of valency with group-number. Valency tested by hydrogen thus shows a maximum of 4; tested by electronegative elements it rises to at least 7, and in OsF$_8$ and probably in OsO_4 to 8. As has been mentioned before (I, p. 14) Mendeleeff first indicated the apparently significant fact that to the right of and including the fourth group the valencies add up to 8. It will also be noticed that in the region to the left of Group IV of constant valency to both types of test element we are

Group	Valency in hydride	Valency in halide or oxide	Numerical total of both valencies
0	0	0	—
I	1	1	—
II	2	2	—
III	?	3	—
IV	4	4	8
V	3	5	8
VI	2	6	8
VII	1	7	8

dealing predominantly with ionized compounds, as it is now certain that hydrides such as LiH, NaH and CaH_2 have crystal lattices in every way similar to those of NaCl and $Ca(OH)_2$ respectively. The only exception is boron, whose hydrides are anomalous and continue to be perplexing, but whose halides BF_3 and BCl_3, as well as the acid $B(OH)_3$, show a clear tervalency.

Chapter III

SOME VALENCY PROBLEMS

SECTION I

It is only in the case of a very few special types of compounds that the bond-diagram is self-evident; it is from such compounds that we obtain our most certain indication of valency (II, p. 16). In the majority of compounds the valency exerted by the constituent elements can only be settled when a thorough examination has led to a definite decision about molecular structure. For example, Kekulé, to maintain the tervalency of phosphorus, proposed the structure $P{<}^{OCl}_{\substack{Cl\\Cl}}$ for phosphorus oxychloride; an alternative structure is $O{=}P{<}^{Cl}_{\substack{Cl\\Cl}}$ in which we appear to accept quinquevalency for phosphorus. If it could once be certainly demonstrated that this compound contains three identical P—Cl bonds, a decision would be made for the second structure and quinquevalency (p. 75). Formerly such examinations had to be accomplished by the methods of pure chemistry, which included a study of reactions reinforced when possible by the application of the principles of stereochemistry (see (1) below). To-day the chemist can draw upon a large array of most powerful physical methods, the operation of which will be briefly explained.* In what follows h = Planck's constant (see p. 106 and Appendix).

(1) *Classical chemical and stereochemical methods*

The classical methods are based largely on the simple idea of comparing the number of isomers theoretically derivable from a proposed molecular structure with the number actually obtainable. For example, if the directions of the four bonds of the carbon atom lay in a plane, or the elements attached all lay in one plane, we should obtain for the molecule CH_2Cl_2 the two isomers shown in plan thus:

$$\begin{array}{ccc} Cl & & Cl \\ | & & | \\ H-C-H & & Cl-C-H \\ | & & | \\ Cl & & H \end{array}$$

* Cf. Sidgwick and Bowen, *Ann. Reports*, 1931, p. 367.

On the contrary, if the bonds were directed from a central carbon atom towards the corners of a tetrahedron (fig. 1, D), only one form of CH_2Cl_2 can arise, as is found in practice. In fig. 1 are drawn the arrangements proposed at various times for the configuration of the five bonds of the quinquevalent nitrogen atom.

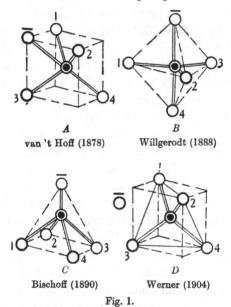

A
van 't Hoff (1878)

B
Willgerodt (1888)

C
Bischoff (1890)

D
Werner (1904)

Fig. 1.

All of these arrangements except the last predict non-existent isomers. From A and B we should obtain two isomers even of the form NR_3HX, i.e. by merely forming a salt of the acid HX from the amine NR_3. Numerous isomers of the form NabcdX would be predicted from all the models A, B and C, but it is an established experimental fact that the same compound is formed in whatever order the groups a, b, c and d are substituted. Model D, which alone introduces the novel idea of not assigning a particular direction to the ionizable bond —X, is analogous to the tetrahedral model for carbon. It is the only model that does not predict non-existent isomers, and was finally confirmed by the X-ray analysis of crystals of $N(CH_3)_4Cl$.* Isomerism of the kind above mentioned arises from differences in interatomic dis-

* Wyckoff, Z. Krist. **67**, 91 (1928); see below, p. 34.

tances within the molecule: this type we may term *positional* isomerism. Positional isomers exhibit distinct physical properties (m.p., b.p., etc.) and usually differ chemically.

Far-reaching results were obtained in the field of *optical isomerism*. A transparent medium is said to be optically active if plane-polarized light, on passing through it, suffers a rotation of the plane of polarization. Optically active media are known in all states—crystalline solids; liquids, including solutions; and gaseous compounds. The optical activity of liquid substances either pure or dissolved in non-active solvents provides the greatest chemical interest, for in such solutions the activity must be an inherent property of the dissolved molecules, and exist independently of their orientation to the light beam. It is recognized that if a certain compound exhibits activity by rotating the plane of polarization, say, clockwise (*dextro*-rotation), it will always be possible to prepare a second 'optical' isomer or *enantiomeride*, of exactly identical chemical and physical properties except that it rotates the plane anti-clockwise (*laevo*-rotation). Usually a moderate stimulus (e.g. some rise of temperature) leads to loss of activity by spontaneous 'racemization', that is, the production from a *d*- or *l*-form of the *d-l* mixture containing exactly equal amounts of the two forms. For the practical methods of separating pure isomers from the *d-l* mixture text-books of organic chemistry may be consulted.

A beam of plane-polarized light can be regarded as composed of two equal superimposed beams of light circularly-polarized in opposite senses, and in like phase. The plane of polarization must rotate if, by differential interaction with the two components of opposite signs, the medium causes a slight retardation of one in respect to the other: in other words, the index of refraction is slightly different for the two circular components. (Differences in the refractive indices of the order of 0·0001 per cent are sufficient to cause marked rotation.) Consider the (not necessarily regular) tetrahedral structure (1) and the reflected image (2) (fig. 2). If the latter is turned round an axis passing through the point *a* so as to bring *d* into the same (right-hand) position as in (1), then it is seen that the positions of *b* and *c* are interchanged (3).

In fact the two types (1) and (2) cannot be superimposed, and are different forms, not, however, owing to *positional* isomerism (see above). Such pairs of related forms so subtly distinguished must always arise from dissymmetrical (including asymmetrical) structures, of which the tetrahedral example above is one of the simplest. Each 'hand' of the dissymmetric structure interacts more with one direction of circularly-polarized light than the other, and the conditions for rotation are realized. Potential or actual optical activity is thus the invariable consequence of dissymmetry in molecular structure.*

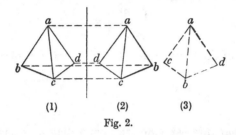

(1)   (2)   (3)

Fig. 2.

It is seen to be essential that the bonds in the dissymmetrical structure should be permanent. Thus if one of the groups a, b, c in the structure

$$\underset{a \;\; b \;\; c}{\overset{X}{\bigwedge}}$$

were ionizable immediate racemization must occur on dissolution in an ionizing solvent, and only a *d-l* mixture will be recovered on evaporation. In the very large number of known optically-active substituted ammonium salts NabcdX, the dissymmetry resides in the grouping Nabcd, and not in the whole group (provided the anion X is not itself a centre of dissymmetry, e.g. anions of active organic acids). Much may therefore be learnt about the nature of bonds from a study of the occurrence of activity.

The tetrahedral configuration required in fig. 2 may obviously be produced (*a*) by placing an effectively *quadrivalent* element at the centre of the figure, (*b*) by placing an effectively *tervalent*

---

* On the use of the terms 'dissymmetric' and 'asymmetric' see Mann and Pope, *Chem. Ind.* **44**, 833 (1925).

element at one of the corners. The optical activity of compounds of carbon, silicon and quadrivalent tin thus supports the tetrahedral distribution of the four bonds issuing from these Group IV elements: the activity of compounds such as ammonium or arsonium salts supports the tetrahedral distribution around the N and As atoms in the cation (see above, and fig. 1), since no positional isomerism is found in all these cases. The activity of the amine and phosphine oxides NabcO and PabcO confirms the tetrahedral model. Cations of the Group VI elements S, Se and Te of the form $(S.abc)^+$ are active, and for these we must suppose the pyramidal arrangement

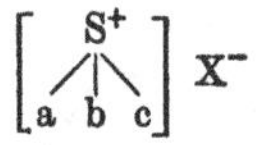

$$\left[\begin{array}{c} S^+ \\ \diagup \vert \diagdown \\ a \quad b \quad c \end{array} \right] X^-$$

The activity of the sulphoxides (1) and esters of sulphinic acids (2)

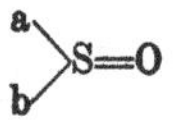

$$\begin{array}{cc} \begin{array}{c} a \\ \diagdown \\ b \end{array} \!\! S\!\!=\!\!0 & \begin{array}{c} a\!\!-\!\!S\!\!-\!\!OR \\ \parallel \\ O \end{array} \\ (1) & (2) \end{array}$$

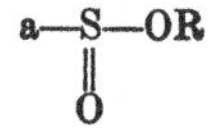

is of particular interest, since the bond-diagrams shown above, if derived from a *quadrivalent* element, would make all the atoms co-planar. The existence of the activity proves a tetrahedral form (see further, pp. 84, 86, 87).

Werner extensively employed the methods of optical stereochemistry in his investigations of complex ions. For example, he showed that the complex cation $(Co.en_3)^{+++}$, wherein en $=$ $NH_2.CH_2CH_2.NH_2$, exhibits activity, and therefore attributed to it the octahedral structure shown in fig. 3. The activity showed little tendency to be lost by racemization, and Werner concluded that the Co—NH$_2$ bonds must be comparable in strength with those in other optically active molecules such as those of carbon, and thus reopened the whole question of

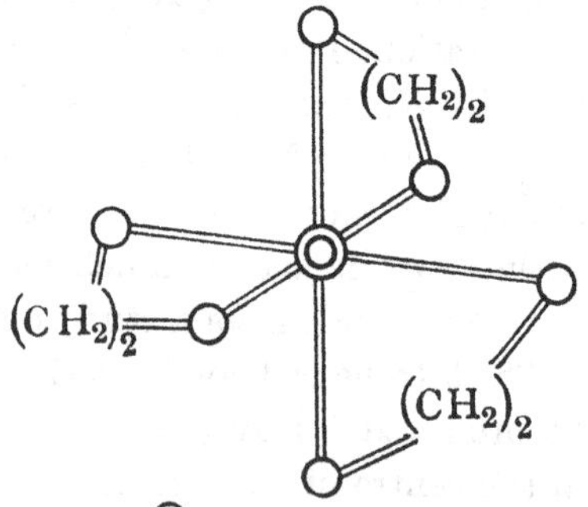

$\circledcirc$, Co; $\circ$, N (NH$_3$)

Fig. 3.

the nature of those substances first termed 'molecular compounds' by Kekulé (I, p. 11).

As further exemplifying the use of stereochemical methods we
may consider the complexes formed by (bivalent) platinum.
Compounds of the general type $Pt.a_2b_2$, e.g. $PtCl_2.2MeNH_2$
are found to exist in two forms, which differ in physical and
chemical behaviour. They are therefore *positional* isomers, the
existence of which immediately disproves a tetrahedral configuration for the complex. Their occurrence would on the other
hand be required if all the four groups lay in a plane (see CH_2Cl_2
above).

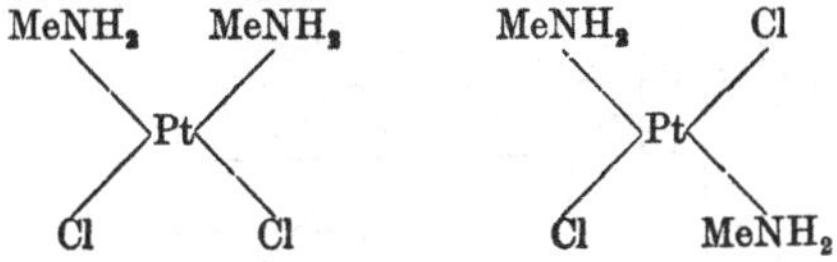

(*cis*-form)   (*trans*-form)

To demonstrate that this is the actual arrangement the compound
shown below was prepared and proved to be optically active:*

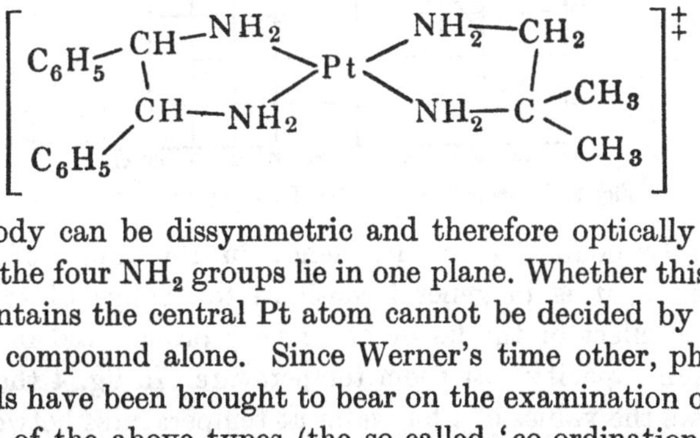

This body can be dissymmetric and therefore optically active
only if the four NH_2 groups lie in one plane. Whether this plane
also contains the central Pt atom cannot be decided by means
of this compound alone. Since Werner's time other, physical,
methods have been brought to bear on the examination of compounds of the above types (the so-called 'co-ordination' compounds), but with the result of extending and confirming
Werner's conceptions rather than of introducing novelties.

(2) *Electric (dipole) moments*

The experimental quantity concerned is the dielectric capacity
ϵ, the fundamental meaning of which may be stated thus: if a
condenser with potential difference E holds, when evacuated, a

---

* Mills and Quibell, *J. Chem. Soc.* 1935, p. 829.

charge q it will hold a charge ϵq for this difference of potential
when a material of dielectric capacity ϵ separates the plates. For
liquids, ϵ ranges from < 5 for hydrocarbons to 80–100 for water
and liquid sulphur dioxide. Dielectric capacity is measured (in
principle) by observing the ratio of the electric capacities of the
same condenser when evacuated, and when filled with the
material concerned. Modern oscillatory circuits containing
thermionic valves are often employed, and advantage taken of
the fact that the (audio-)frequency of such circuits is proportional to $1/\sqrt{(LC)}$, where L is the induction, and C the total capacity. It is found that while the dielectric capacity of liquids is

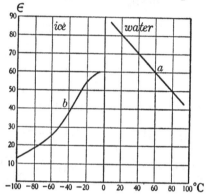

Fig. 4. Dielectric capacity (ϵ) of water (a) and ice (b).

practically unaffected by frequency in the range of audio-
frequencies, it is in general sensitive to change of temperature, the effect of the latter being very pronounced in liquids
with high capacity ϵ at room temperature. In fig. 4 the curve
(a) shows the values of ϵ for water at temperatures between the
freezing and boiling points. The decrease with rise of temperature
is almost linear. On the other hand, the dielectric capacity of ice
is influenced markedly by both frequency and temperature.
Curve (b) shows the dielectric capacity of ice measured with the
low frequency of 120 cycles per sec.

It is now universally accepted that this complex behaviour of
water and other liquids showing similar effects is attributable to
the presence of one or more permanent dipoles, i.e. of equal

amounts of positive and negative electricity separated by finite distances of molecular dimensions. These dipoles tend to be oriented against the electric field applied to determine the dielectric capacity, but the 'randomizing' effect of thermal motion in liquids opposes this setting. The average value of the *effective* molecular dipole, $\bar{\mu}$, is thus determined by three factors: the value of the permanent molecular dipole μ_0 (equal to *er*, where *e* is the resultant charge separated by the resultant distance *r*); the orienting force, which is proportional to the applied electric field strength *F*; and, in opposition to these, the thermal motion. It can be shown that the actual relationship is

$$\bar{\mu} = \frac{\mu_0^2}{3RT} F.$$

It is of the greatest interest to notice that some sort of mobility is retained by the dipoles even in the solid ice, but it is not certain whether the orienting motion here is of the same nature as in liquids.*

Measurement of the dielectric capacity of pure liquid proves the existence of dipoles, but it is not well adapted to afford an accurate estimate of the molecular electric moment (μ_0). In deriving the above relation it is assumed that all the dipoles act independently, but in the close-packed liquid system this cannot be true; even if molecular association (as in water) is absent. Experiments are necessary upon dilute solutions of the substance in a non-polar solvent; benzene is frequently used. In such a manner the electric moments of a very large number of compounds have been fixed. By comparison of suitably related compounds the moments of typical bonds have been computed. Table 4 gives the moments of some commonly occurring bonds (an extensive list of molecular moments will be found in *Trans. Faraday Soc.* **30** (1934), Appendix). Since the dipole charge *e* is of the order of magnitude of the ionic charge $4 \cdot 80 \times 10^{-10}$ e.s.u., and the dipole distance *r* of the order of 10^{-8} cm., a molecular electric moment is of the order

$$er = 10^{-18} \text{ e.s.u.} \times \text{cm.}: 1 \text{ e.s.u.} \times \text{cm.} \times 10^{18} = 1 \text{ debye (D.).}$$

* Cf. Debye, *Polar Molecules*, 1929, p. 102.

Table 4. *Electric moments* *

+→	Moment × 10^{18}	+→	Moment × 10^{18}
H—N	1·31	C=O	2·4
H—O	1·53	N=O	1·9
C—N	0·4	C≡N	3·6
C—O	0·86	—	—
C—Cl	1·56	—	—

* Cf. Smyth, *J. Physical Chem.* **41**, 209 (1937).

Of course, in symmetrical molecules individual moments may be self-neutralizing. The dipole of C—Cl is revealed in CH_3Cl and $CHCl_3$ but completely self-neutralizing in CCl_4; that of C=O gives strong polarity to acetone $(CH_3)_2CO$ but none to the linear molecule CO_2. From Table 4 it may be concluded that polarity is a *normal* property of any bond between *unlike* atoms (*see* IV, *p. 120*).

(3) *Methods depending on the diffraction of high-frequency radiation*

The strides made by the method of X-ray diffraction in elucidating crystal structure are amongst the most remarkable of modern science. In favourable cases a completely defined model of the arrangements of atoms building a crystal can be obtained, always excepting the position of hydrogen (IV, p. 117). Summaries of the experimental technique are now so numerous and well known that it would be supererogatory to recapitulate them here.†

The method of X-ray diffraction is often complicated by the consideration already alluded to under electric moments (2) above; the diffraction pattern is produced not by assemblies of independent molecules or ions, but by dense-packed inter-dependent units. Such a difficulty is inherent in the use of crystals, but would be absent if gases or vapours were used. X-radiation proves insufficiently powerful to treat gases, and it is therefore fortunate that the variation known as 'electron diffraction' has been found workable.

† Bragg, *The Crystalline State*, vol. I; James, *X-Ray Crystallography*, 1953; Bunn, *Chemical Crystallography*, Oxford, 1945; Robertson, *Organic Crystals and Molecules*, Ithaca, New York, 1953.

In the light of modern quantum physics, an electron moving at high speed is regarded as having properties equivalent to high-frequency radiation, and it has the advantage that its associated wave-length $\lambda = h/p$ is adjustable merely by changing the momentum p, i.e. changing the electric field strength through which the electron stream passes. By this modification of the diffraction method it is possible to gain a great deal of information about the geometrical structure of the molecules of volatile substances, but the methods of calculation are usually much less certain than those of the X-ray method for crystals.*

(4) *Methods based on the investigation of infra-red and similar spectra*

Infra-red spectra arise from the rotation and/or intramolecular vibration of polar structures. Both types of motion are quantized, and the general principles applicable to atomic spectra are again of service. The very great complexity of molecular or 'band' spectra imposed on an already very difficult technique hinders their rapid interpretation. In so far as molecular *rotation* is concerned in producing the spectra it is to be expected that a determination of molecular *moments of inertia* will eventuate. From these we may at once, since the masses are known, calculate interatomic distance or *bond-lengths* in sufficiently simple molecules (Table 5). Such distances are also deducible from the diffraction method (2), and the methods are complementary, especially in the sense that the difficulties in 'placing' the lighter atoms, especially hydrogen, arising in the diffraction methods from the low absorption and scattering power of these atoms are of course not encountered in the spectral method. From the *vibrational* motions an estimate can be obtained of the *bond-strength*, which is not accessible to any other method. Bond-strength is suitably measured by the so-called *force constant*, which is numerically equal to the force required to stretch the bond along its axis through unit distance.

* For a detailed description of the method and typical results see Brockway, *Rev. Mod. Phys.* 8, 231 (1936), and Allen and Sutton, *Acta Cryst.* 3, 46 (1950). For an account of neutron diffraction see Thewlis, *Ann. Reports*, 47, 420 (1950).

Some general results of the application of modern physical methods

(1) *Molecular individuality in the solid state*

Some of the results of crystal analysis have proved rather disconcerting to previous conceptions about the units out of which the crystal is built. In the form of silica known as β-cristobalite it has been found that the arrangement of the constituent Si and O atoms is as follows: around each Si atom are four equidistant and tetrahedrally disposed O atoms; while each O atom has two Si atoms as its nearest neighbours equidistant from it (consult fig. 5). The other forms of silica have similar dispositions of the

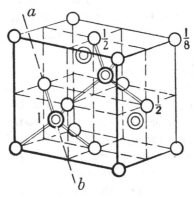

Fig. 5. The cubical unit cell of zinc blende (ZnS) and diamond. ⊚, S; ◯, Zn. For diamond S = Zn = C. (For SiO_2 put C = Si, and interpose O at midpoint between each pair of Si atoms.)

atoms with only slightly less symmetry. No evidence is to be found for the existence of the supposed molecule SiO_2. If we ask what other evidence we have for the existence of this molecule, it will be admitted that the principal item is the chemical analogy with CO_2. Certainly no gaseous molecule SiO_2 has been proved to exist. It must be conceded that the chemical union of silicon and oxygen produces a crystalline solid, of *analytical composition* represented by the formula SiO_2, but having the atomic arrangement outlined above. In zinc blende and wurtzite (ZnS) and zinc oxide, each Zn atom is tetrahedrally surrounded by four sulphur (or oxygen) atoms, and each sulphur (or oxygen) by four zinc atoms (figs. 6 and 6 a). The crystal schemes of silicon carbide SiC (carborundum) and of silver iodide (at room temperature) are exactly similar. In the numerous cases of which these are typical examples we have to admit that the bond-diagram has infinite extent, corresponding to a 'macromolecule'.

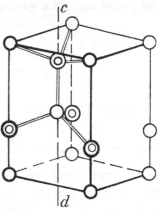

Fig. 6. The hexagonal unit cell of wurtzite (ZnS).
◎, S; ○, Zn.

On the other hand, while the crystal structures of carbon (diamond and graphite, p. 197), silicon, and probably boron, show macromolecules, the crystals of other elementary substances provide confirmation of older chemical suppositions. Solid oxygen, nitrogen and iodine are, for example, formed from the actual molecules O_2, N_2 and I_2: for in the crystal the atoms occur in 'close' pairs. The volatility and, for oxygen and

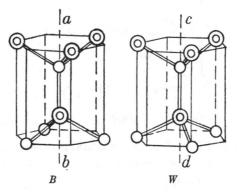

B   W

Fig. 6 a. The closely-related structures of zinc blende (B), and wurtzite (W).
◎, S; ○, Zn. Axes ab and cd are shown in figs. 5 and 6.

nitrogen, the low melting-point are due to weak non-chemical (van der Waals') forces attaching the molecules together. In phosphorus we find P_4,* and in sulphur S_8 (figs. 7, 8). The mutability of this last large molecule into long-chain or 'fibre' molecules accounts for the previously perplexing nature of 'plastic' sulphur.† Crystals of organic substances are in general built of the actual molecules envisaged by the organic chemist.

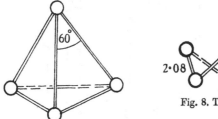

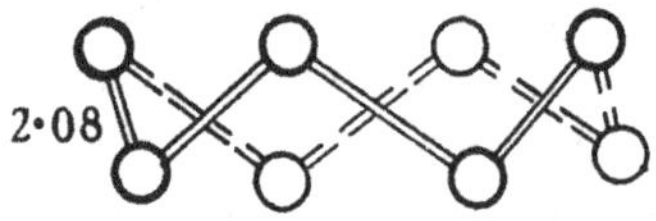

Fig. 8. The structure of S_8.

Fig. 7. The structure of P_4 and As_4.

The structures of ZnS, ZnO, the sulphide and oxide of a B-family metal (pp. 24, 27), are sharply contrasted with those of the sulphides and oxides of the corresponding A-family, Mg, Ca, Sr and Ba, the lattices of which are dense-packed and consist wholly of

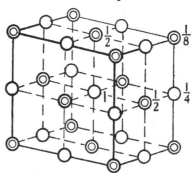

Fig. 9. The cubical unit cell of NaCl or MgO. ◎, Cl, O; ○, Na, Mg.

ions (M^{++} and $S^=$ or $O^=$); the structural plan is exactly like that of NaCl, namely the surrounding of each anion or kation by *six* equidistant ions of opposite charge (fig. 9). Again no molecule

* The molecules P_4, As_4, cyclopropane C_3H_6 and ethylene oxide $CH_2{-}CH_2$ / $\backslash O /$ are examples of the small group of molecules exhibiting bonds as close together as 60° (see p. 141).

† Gingrich, *J. Chem. Physics*, 8, 29 (1940); Bacon and Fanelli, *J. Amer. Chem. Soc.* 65, 639 (1943); Powell and Eyring, *idem*, 648.

such as NaCl or MgO is discernible. In the crystals of salts of the oxy-acids (sulphates, nitrates, carbonates) an ionic lattice is again discovered, but the anions consist of compact assemblies such as SO_4, NO_3 and CO_3, in which there is no doubt that the constituent atoms are in firm chemical bonding. If we are to point to a molecule in such crystals, it would be to these charged groups, and not for example to $CaCO_3$.

A knowledge of the ionic arrangement in the crystal renders possible the computation of the lattice-energy, i.e. the (largely coulombic) energy released on assembling the prepared ions into one g.-mol. of crystal structure. The following are typical lattice energies:*

	F	Cl	Br	I	
Na	222	183	172	159	Cal.
K	192	164	156	144	Cal.

The lattice-energy is a fundamental quantity for a crystal, and from it many of the physical properties of the crystal can be evaluated. Conversely, a lattice may be proved to be ionic by agreement between such calculations and actual properties (for example, see this chapter, p. 102, MnO_2 and cited ref.).

Other typical ionic lattices are fluorite, CaF_2 (Fig. 10), and caesium chloride, CsCl (Fig. 11).

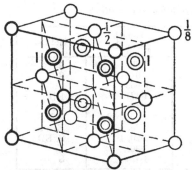

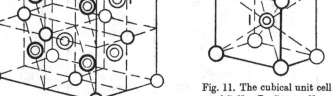

Fig. 11. The cubical unit cell of CsCl. ◎, Cs; ○, Cl.

Fig. 10. The cubical unit cell of fluorite (CaF_2) or Li_2O.
◎, F; ○, Ca.   ◎, Li; ○, O ('anti-fluorite' structure).

* Evans, *Crystal Chemistry*, 1939, p. 57; Born, *Verh. dtsch. phys. Ges.* **21**, 679 (1919); Mayer and Helmholz, *Z. Physik*, **75**, 19 (1932).

Note on the assessment of the number of molecules contained in a unit cell

It must not be forgotten that the position of an ion (or group) in the unit cell determines for it a chemical 'value', which is fractional unless the ion or group is completely enclosed within the cell.

Position	Number of cells to which ion is common	Value in unit cell
Corner	8	$\frac{1}{8}$
Edge	4	$\frac{1}{4}$
Face	2	$\frac{1}{2}$
Within cell	1	1

Examples:
$$Cl^- = (8 \times \tfrac{1}{8}) + (6 \times \tfrac{1}{2}) = 4,$$
$$Na^+ = (12 \times \tfrac{1}{4}) + 1 = 4.$$

Hence NaCl has four molecules to its unit cell (fig. 9).
$$Ca = (8 \times \tfrac{1}{8}) + (6 \times \tfrac{1}{2}) = 4,$$
$$F = 8.$$

Hence CaF_2 also has four molecules to its unit cell (fig. 10).

(2) *Bond-lengths*

By bond-length is meant the distance (usually expressed in Angstrom units = 10^{-8} cm.) between the nuclei of the bonded atoms. In Table 5 will be found a list of typical bond-lengths, most of which are established by the complementary application of several physical methods.

Table 5. *Typical bond-lengths*

Bond	Compound giving length	Length (A. units)	Method
C—H	CH_4, etc.	1·093	S
N—H	NH_3	1·016	S
O—H	H_2O	0·955	S
C—N	$(CH_3)_3N$	1·47	ED
C—O	$(CH_3)_2O$	1·42	ED
C—F	CF_4, CH_3F	1·36, 1·39	ED, S
C—Cl	CCl_4, CH_3Cl	1·75, 1·77	ED, S
C=O	H_2CO, $COCl_2$	1·20, 1·17	S, ED
C=N	(See p. 215)	1·30	X
C≡N	HCN	1·154	S

Methods: S, spectroscopic; ED, electron diffraction; X, X-ray (crystal) diffraction.

One of the most striking features is the closer approach of the atomic centres in a multiple bond, which Kekulé prophetically described as a 'more compact' form of linking (p. 11). It will be obvious that in this characteristic shortening we have a very valuable clue to the presence of multiple bonds in a molecular structure, and one that is independent of chemical evidence on the point (see, for example, this chapter, p. 73).

It appears probable that bond-lengths, more especially of single bonds, may be considered as the sum of standard and in-variable atomic radii ('covalent' radii, below). Undoubtedly the best established bond-length is C—C, which is constant at 1·54 A. in diamond, and a large range of non-aromatic hydrocarbons (paraffins and cycloparaffins). Using as a datum the radius of C— as $1·54/2 = 0·77$, we can now calculate the radii of other elements X by deducting C— from C—X. The cross-checking of values so obtained is reasonably satisfactory. It must however be remembered that the product of the interaction of molecules A–A and B–B, in which B is the more electronegative atom, is not A–B but A_+–B_-, in which the dipole strength increases with the difference in electronegativity (see Table 4). For this reason we may expect, and indeed find, that deviations from the additivity of bond lengths are most serious in links between atoms of widely different electronegativity: for example in the links between most other atoms and fluorine.* In Table 6 the steady decrease in radius in the series C to F and Si to Cl is to be noted.

Table 6. *Radii of combined atoms* ('*covalent radii*')

H—	Hydrides	0·30
C—	[Diamond]	0·771
C=	C_2H_4	0·665
C≡	C_2H_2	0·602
N—	N_2H_4	0·735
N=	$(C_6H_5)_2N_2$	0·60
N≡	N_2	0·547
O—	H_2O_2, $S_2O_8^-$	0·74
O=	O_2	0·604
F—	CF_4, CH_3F	0·59–0·62
Si—	Si_2H_6	1·16
P—	P_4	1·10
S—	S_8	1·04
Cl—	Cl_2	0·99

* For a discussion of the principle of additive atomic radii see Scho-maker and Stevenson, *J. Amer. Chem. Soc.* **63**, 37 (1941), and Rogers, Schomaker and Stevenson, *J. Amer. Chem. Soc.* **63**, 2610 (1941).

(3) *The shape of molecules*

Every molecular structure involving more than one atom has in principle three moments of inertia about three spatial directions, intersecting in the centre of gravity and mutually at right angles. It is convenient to choose as the three directions, or axes of rotation, the symmetry axes of the molecule (fig. 12). According to the degree of symmetry of the molecule all these moments of inertia may be equal, two may be equal, or all three unequal. For brevity we may say that the first class has one moment, the second two, and the last and least symmetrical three moments of inertia. A molecule in which all the atoms lie in a line (linear structure) has in principle the axis of one moment along the molecular axis, and passing through the atomic centres. The moment in this case involves only electronic masses, and is therefore vanishingly small. According to modern quantum theory the quantum of rotational energy ($h^2/8\pi^2 I$) is inversely proportional to the moment of inertia I, and this mode of rotation could hence only be excited under such drastic conditions as are unlikely to be reached experimentally. The same considerations preclude the rotation of a monatomic molecule.

An investigation of the infra-red spectrum of simple molecules allows both the number and the values of the moments of inertia to be accurately determined. The number of moments enables the symmetry type and therefore the shape of the molecule to be settled; and from the numerical values of the moments we draw our most accurate values of bond-lengths (see p. 44). The existence or absence of an electric moment may also greatly elucidate the question of molecular shape. Thus from the fact that water has a strong electric moment we deduce that it cannot be a linear molecule (HOH); the absence of an electric moment in CCl_4 and its presence in CH_3Cl supports a regular tetrahedral form for CCl_4, in which the four individual dipoles are self-neutralizing. A combination of knowledge of bond-length with that of electric moment may often throw light on the constitution of a molecule. Thus if the full ionic charges $e = 4 \cdot 80 \times 10^{-10}$ e.s.u. were placed at the positions of H and Cl in HCl, the expected moment would be $4 \cdot 80 \times 10^{-10} \times 1 \cdot 27 \times 10^{-8} = 6 \cdot 10 \times 10^{-18}$

e.s.u. × cm., since the bond-length is 1·27 A. The actual moment is only $1·03 \times 10^{-18}$, and we deduce that HCl is not to be regarded as an ion-pair. The molecules CO_2, N_2O, HCN and C_2H_2 have only one moment of inertia, and are *linear*. The values of the moments of inertia, and other details of the spectra, show that CO_2 is OCO, while N_2O is NNO, and not NON as formerly seemed more probable. Methane CH_4 also has only one moment of inertia, and must therefore be in the shape of a regular *tetrahedron*, as in the classical conception of van 't Hoff and Le Bel. The hydrides NH_3 and PH_3 are of *pyramidal* shape, with N and P at one apex, since they exhibit two moments of inertia. H_2O, SO_2 and NO_2 all have three distinct moments, and are therefore *triangular* in shape (fig. 12).

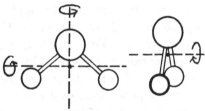

Fig. 12. Rotational axes of the molecule AB_2.

The method of electron-diffraction provides our knowledge of the shapes of many gaseous halides. In BF_3 all four atoms lie in one plane, the halogen atoms being at the corners of an equilateral triangle, with boron at its centre. The shape of this molecule is thus sharply contrasted with the pyramidal NH_3

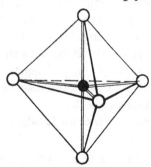

Fig. 13. The structure of PF_5.
●, P; ○, F.

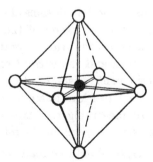

Fig. 14. The structure of SF_6.
●, S; ○, F.

and PCl_3. In PF_5 the phosphorus lies at the centre of a regular trigonal bipyramid (fig. 13), and all P—F bonds are equal in length. We may contrast this simple structure with the complex *crystal* arrangement of PCl_5.* It seems that here there are two units, PCl_4 and PCl_6, and eight of one unit surround one of the other, as in the ionic lattice of CsCl (fig. 11). It is therefore considered that the units are ions: PCl_4^+ and PCl_6^-. In SF_6 all the S—F bonds are equal in length, and the F atoms are at the corners of a regular octahedron (fig. 14).

SECTION II

We may now proceed to illustrate the application of both chemical and physical methods to selected problems classified under the Groups of the Periodic System. In this chapter we shall continue to regard valency as defined by classical conceptions already discussed in Chapter II. In assigning valencies to elements contained in ions we shall continue to make no distinction between bonds capable of ionization and other bonds. Thus we shall regard oxygen as bivalent in the ions $O^=$, O^-—O^- (in peroxides) and OH^-, as well as in the compounds H_2O and CH_3OH. Nitrogen will be taken as quinquevalent in both NH_4^+ and NO_3^-.

It is intended in this Section to take the elucidation of structure as far as modern experimental methods will permit, but to defer presenting the solutions of the numerous problems so raised until after the discussion of the electronic theory in later chapters. It is hoped that this separation will more clearly define both the problem and the current explanation, and by emphasizing the permanent insistence of the former and the power (and defects) of the latter lead to a clearer understanding of both.

In presenting the data upon molecular structure it will be convenient to adopt the following contractions:

'A, B' should be read 'the distance, in Angstrom units (10^{-8}

---

* Clark, Powell and Wells, *J. Chem. Soc.* 1942, p. 642.

cm.), between the nuclei of A and B (in a compound containing them)':

'∠ABC' means the interbond angle subtended at the centre of atom B by the centres of atoms A and C, however these are bonded to B.

GROUP I (A)—THE ALKALI METALS

(1) With the unimportant exception of the elementary molecules, e.g. Na_2 (of which there is a minute proportion in the vapours), and a few organo-metallic compounds, e.g. of lithium (p. 51) all the compounds of these metals are ionic in type. In addition to the evidence of general physical properties—high melting-point, low volatility, solubility in water, and low solubility in organic liquids—we have for a very large number of these compounds the direct confirmation of X-ray analysis (p. 43). This group of compounds therefore forms an excellent field for examining the chemical effects related to ionic charge and size (cf. IV, p. 117). The ionic radii, obtained from crystal measurements, of the Group I (A) and Group II (A) metals, are recorded in Table 7.

Table 7

	Li^+	Na^+	K^+	Rb^+	Cs^+	Mg^{++}	Ca^{++}	Sr^{++}	Ba^{++}
r	0·67	0·98	1·33	1·48	1·65	0·78	1·06	1·27	1·43
r^2	0·45	0·96	1·77	2·20	2·73	0·61	1·12	1·62	2·05

We may plausibly expect the effects of cations upon their anionic neighbours to be measured by the superficial charge density, proportional to z/r^2, where z is the ionic charge.

To exemplify the effects we may consider the facts about the thermal stability of certain salts, summarized below:

Carbonates: those of Na, K, Rb and Cs are not decomposed. From the pressure of CO_2 at a given temperature the order of stability of other carbonates is as follows:

$$Ba > Sr > (Li, Ca) > Mg.$$

Nitrates: those of Na, K, Rb and Cs yield nitrites; those of Li and the Group II metals all yield oxides.

In these thermal changes the reaction involved can be stated thus:

$$M^+ \text{ (or } M^{++}) + XO_3^- = M_2^+ . O^- \text{ (or } M^{++}. O^=) + XO_2,$$

and it is clear that the power of the cation to extract the ion O^- from the anion XO_3^- exactly follows the values of z/r^2.

(2) *The oxides of the alkali metals*

The most stable oxides, judged by the results of heating the metal in oxygen, are as follows:

Li_2O (Li_2O_2 is formed only by the interaction of LiOH and H_2O_2),

Na_2O_2 colourless,

KO_2, RbO_2, CsO_2 all deep yellow.

The *superoxide* KO_2 possesses the (tetragonal) crystal structure of CaC_2, in which the ion O_2^- replaces C_2^-, and is the source of the colour (p. 58). Here again we may trace the effect of ionic size. Successive addition of oxygen must proceed thus:

$$O^= + \tfrac{1}{2}O_2 \rightarrow O^- \!-\! O^-; \qquad O^- \!-\! O^- + O_2 \rightarrow 2O_2^-$$

as in	as in	as in
R_2O	R_2O_2	RO_2

If the cation is small and powerful it will, so to speak, anchor the electrons in the oxygen, tending to retain as the most stable grouping $R_2^+O^=$, and hindering or preventing the progressive spreading and weakening of the anionic charge over more oxygen atoms. It is to be noted that while peroxides RO_2 of Group II metals are well known, no superoxides have been discovered in this group.

(3) *The polyiodides of the alkali metals*

In the formation of polyiodides the necessary reaction is of the type

$$I^- + I_2 \rightarrow I_3^-, \text{ etc.,}$$

which like the production of higher oxygen ions is a process of spreading of the negative charge. It is therefore not unexpected to find that the relationships of the polyiodides of the alkali metals closely resemble those of the oxides.

Comparison of the polyiodides

Li and Na: no solid iodide of the above type has been isolated.*

K: the hydrate $KI_3.H_2O$ is stable at 25°. Dehydration is accompanied by loss of iodine.

Rb: RbI_3 is readily prepared by adding I_2 to an aqueous solution of RbI. It may be recrystallized from either water or alcohol.

Cs: CsI_3 is prepared by adding only the theoretical amount of I_2 to an aqueous solution of CsI.

Dissociation pressures of iodine

$t°$	160	180	200	
p (mm.)	{300	510	700	RbI_3
	{ 20	70	160	CsI_3

When an excess of iodine is added to a solution of CsI the stable CsI_4 separates as very well-formed crystals.

(4) Hydrides and organo-metallic compounds

Lithium hydride, LiH, has been very fully examined. Formed by direct union, it has the crystal structure of LiCl (cubic), and yields hydrogen at the anode when fused and electrolysed. There is very little doubt that the hydrides of the other alkali metals are analogous and that they all contain the anion H⁻.

Numerous organo-metallic compounds are known, more particularly those of Li and Na, which are colourless solids, insoluble in benzene and other hydrocarbon solvents. They must all be regarded as ionic, with a negative alkyl ion replacing H⁻ in the hydride RH.

Alkyls of sodium. $NaCH_3$, NaC_2H_5, NaC_3H_7, NaC_8H_{17} and NaC_6H_5 are all colourless solids. The two partly aromatic compounds sodium benzyl, $NaCH_2.C_6H_5$, and sodium triphenylmethyl, $NaC(C_6H_5)_3$, are deep-red in colour, insoluble in hydrocarbons, and yield conducting solutions in ether.†

Alkyls of lithium. $LiCH_3$ is a colourless crystalline solid, resembling the lower sodium alkyls. LiC_2H_5, m.p. 95°, is soluble in benzene, and LiC_3H_7 is liquid at room temperature and mixes

---

* Briggs, Geigle and Eaton, *J. Physical Chem.* 45, 595 (1941).

† Schlenk and Marcus, *Ber.* 47, 1665 (1914).

in all proportions with benzene. These last two compounds are clearly not ionic, but must be regarded as organic derivatives of ethane and propane respectively.

(5) *The amides of the alkali metals* ($NaNH_2$)

Sodamide, produced by the direct action of NH_3 upon the metal, forms transparent crystals, m.p. 208°, of which the structure has been shown (by X-ray analysis) to consist of Na^+ and NH_2^- ions*. The following thermal equations are of interest:

$$NH_3 \quad + Na = NaNH_2 \quad + \tfrac{1}{2}H_2 + 20 \cdot 86 \text{ Cal.}$$
$$H_2O(g) + Na = NaOH \quad + \tfrac{1}{2}H_2 + 25 \cdot 40 \text{ Cal.}$$
$$H_2O(g) + Na = Na^+ . OH^- + \tfrac{1}{2}H_2 + \text{ca. } 30 \text{ Cal.}$$

The last equation has been computed on the assumption that the lattice energy of NaOH is about the same as that of NaF (p. 43). Sodamide conducts electricity in the fused state, the conductance being comparable with that of NaOH at the same temperature. These data seem to confirm that $NaNH_2$ is to be regarded as $Na^+ . NH_2^-$.

GROUP I (B)

(1) *The valency of copper*

Familiarity with the oxy-acid salts of copper, such as the sulphate, is apt to give the impression that compounds in which copper is bivalent are more stable than the cuprous series. The following facts suggest the contrary, which would be in accordance with the demands of the periodic system. The halides $CuCl_2$ and $CuBr_2$ lose halogen on heating, the bromide at a lower temperature than the chloride. The iodide CuI_2 appears to be unstable under all conditions. CuO dissociates to Cu_2O and oxygen on strong heating, and $Cu(CN)_2$ yields CuCN on gentle warming. The widespread occurrence of Cu_2S and Cu_2O in contrast with the sparseness of CuS and CuO is clearly explained by the superior stability of the former ores.

The cuprous salts of oxy-acids, which can readily be prepared in non-aqueous media, yield a precipitate of copper and the cupric salt on contact with water. This behaviour, with few

---

* Juza, Weber, and Opp, *Z. anorg. Chem.* **284**, 73 (1956).

parallels among metallic compounds, may be explained if the reaction is viewed, not as $2Cu^+ = Cu^{++} + Cu$, but as

$$2Cu^+ \cdot aq. = Cu^{++} \cdot aq. + Cu,$$

in which we explicitly recognize the hydration of cations in aqueous solution. To remove one electron from each of the anhydrous and free entities Cu and Cu^+ needs the input of 180 and 285 Cal./g.-atom respectively. On the contrary, the energy released on the hydration of Cu^+ (~ 120 Cal./g.-atom) is much less than that set free when Cu^{2+} is hydrated (~ 530 Cal./g.-atom). Hence the first, anhydrous reaction would be highly endothermic, while the second, proceeding in water, is strongly exothermic. This example emphasizes the essential part played by hydration in the reactions of ions.

Of the cupric compounds, oxy-salts, such as the sulphate and nitrate, are commonly ionic, but the deep-brown anhydrous halides $CuCl_2$* and $CuBr_2$,† and the black cupric oxide are macromolecular in the crystalline state. The cuprous halides are all colourless, and show the zinc blende structure (p. 41), which (or the closely related wurtzite structure) is found only in binary compounds in

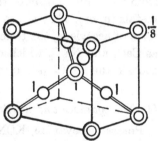

Fig. 15. The cubical unit cell of cuprite, Cu_2O (or Ag_2O). O, Cu; ◎, O.

which the atoms are both drawn from Group IV (e.g. CSi) or one from each of two groups equidistant from Group IV (e.g. BeO, ZnS, AlN, AlAs). In the structure of cuprous oxide oxygen atoms are surrounded tetrahedrally by four copper atoms, and copper atoms lie midway on the lines joining oxygen atoms (fig. 15).

(2) The halides of silver: silver oxide

Contrary to expectation only AgI shows the zinc-blende structure, AgCl and AgBr both possessing the NaCl structure and thus being ionic. Mixed crystals of the iodide and bromide

---

* Wells, J. Chem. Soc. 1947, 1670.
† Helmholz, J. Amer. Chem. Soc. 69, 806 (1947).

containing up to 50% of each are readily formed, and as a whole possess the NaCl structure.

Silver oxide, Ag_2O, has the same unusual crystal structure as Cu_2O (fig. 15), but dissolves readily in oxy-acids and, unlike Cu_2O, exhibits in such reactions no disproportionation.

(3) Acetylene derivatives of copper and silver

A red insoluble solid, of composition Cu_2C_2, and a colourless insoluble solid, of composition Ag_2C_2, are precipitated by the action of acetylene upon ammoniacal solutions of cuprous and silver salts respectively. By their indifference to water, their explosive nature when dry, and by the colour in the case of the copper compound, these substances are sharply distinguished from the *carbides* of both the alkali metals and of the Group II metals (p. 58). They are to be regarded as *acetylides*, i.e. derivatives of acetylene, the term *carbide* being reserved for *salts* such as CaC_2 and Na_2C_2 which can be shown by crystal analysis to contain the ion C_2^- (see further on acetylides, *p. 202*).

(4) The cyanides and nitrites of Group I metals

Potassium cyanide, KCN, does not crystallize, as it might be expected to do, in a tetragonal lattice similar to that of calcium carbide (fig. 19, p. 58), but in a lattice of cubical symmetry. The cubical unit cell, of NaCl structure (p. 42), has an edge (6·55) almost equal to that of KBr (6·57), which also has the NaCl structure. For the linear group —CN to attain cubical symmetry it must be assumed to rotate round axes drawn at random through the centre of its lattice position, and so to occupy a spherical space equal in radius to that of a bromine ion. Silver cyanide exhibits, on the contrary, a non-ionic lattice, with a rhombohedral unit cell (fig. 16). The crystal is composed of indefinitely long chains —C—N—Ag—C—N—Ag— arranged in hexagonal packing. The Ag, Ag distance of 5·26 is too short for an ionic lattice.*

The crystallographic distinction between $NaNO_2$ and $AgNO_2$

* West, Z. Krist. **88**, 173 (1934) and **90**, 555 (1935).

is not so obvious, but no less significant.* The edges of the orthorhombic unit cells of both compounds show the configuration

$$M(1)\frac{O(1)}{O(2)}N...M(2), \quad \text{where } M = \text{the metal (fig. 17),}$$

with the following distances:

Na(1), N = Na(2), N = 2·78,
Na(1), O(1) or O(2) = 2·46,
Ag(1), N = 3·12; Ag(2), N = 2·04,
Ag(1) or Ag(2), O(1) or O(2) = 2·69.

The sum of the accepted (covalent) radii of Ag (1·44) and —N (0·735) is 2·175. It thus seems probable that the lattice of $NaNO_2$ is purely ionic, while that of $AgNO_2$ is formed of molecules of 'nitro-silver'. The production of iso- nitriles by the action of AgCN upon alkyl halides, and of nitro-compounds

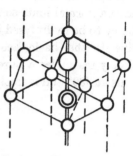

Fig. 16. The rhombohedral unit cell of AgCN.

◎, C; ○, N; ○, Ag

(vertical broken lines show direction of —CN—Ag—CN— chains).

by the action of $AgNO_2$ upon alkyl halides, is in obvious agreement with these crystallographic data.

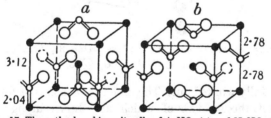

Fig. 17. The orthorhombic unit cells of $AgNO_2$ (a) and $NaNO_2$ (b).
○, N; ○, O; ●, Ag or Na.

GROUP II

(1) Beryllium

Although the constant bivalency of this element associates it
with other Group II metals, it differs sharply from those in the
A family (Mg, Ca, Sr and Ba) in its tendency to form non-ionic
compounds.

The wurtzite (macromolecular) structure of BeO resembles
that of ZnO (p. 41). The chloride $BeCl_2$ fumes in moist air

* Ketelaar, Z. Krist. **95**, 383 (1936).

like PCl_5, has a m.p. 400°, fails to conduct electricity when fused, and readily sublimes. The absence of electric moment indicates a linear structure Cl—Be—Cl. The alkyl $Be(C_2H_5)_2$ is a liquid, b.p. 186°. Salts of the oxy-acids—sulphate, nitrate, carbonate—have a normal ionic constitution, but show a very marked tendency to be hydrolysed to basic salts, of which the most remarkable is the basic acetate, $Be_4O(CH_3CO_2)_6$. This substance closely resembles a purely organic compound; it melts at 283°, boils at 330° and is freely soluble in $CHCl_3$. Its crystal structure

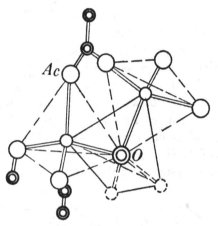

Fig. 18. The structure of beryllium basic acetate $Be_4O(CH_3CO_2)_6$.
O, Be; O, O of CH_3CO_2; ⊙, C.

has been fully worked out. The molecule consists of four nearly regular tetrahedra, two of which are shown in fig. 18. The tetrahedral relation of the four oxygen atoms (Ac and O, fig. 18) to the central Be is that of BeO. Three of the oxygen atoms (Ac) are derived from the acetyl groups, and the fourth (O) is the atom of the formula above. This last atom links the four tetrahedra together by the corners; further links are provided by the 'bridge' O—C—O from the acetyl groups. Not the least interesting of the results of this analysis is that the oxygen atoms of the acetate group CH_3CO_2 act in an equivalent manner (*p. 210*).

In connection with the strong tendency of Be to form non-ionic (i.e. non-metallic) compounds it should be noted that Be^{++} is by far the smallest metallic ion ($r = 0\cdot34$ A.), and therefore possesses very great deforming power (cf. IV, *p. 117*).

(2) *Magnesium—Grignard reagents*

The relative ease of their preparation and especially their non-inflammability have together been the chief reasons why since their discovery in 1900 these reagents have replaced the formerly used metallic alkyls, such as $Zn(C_2H_5)_2$, in synthetic organic chemistry. Structural formulae that have from time to time been in favour, such as

$$C_2H_5 \diagdown O \diagup Mg(hal.) \qquad \text{or} \qquad (C_2H_5)_2O \diagdown Mg \diagup (hal.)$$
$$C_2H_5 \diagup \quad \diagdown (alkyl) \qquad\qquad (C_2H_5)_2O \diagup \quad \diagdown (alkyl)$$

so obviously raise questions of valency that Grignard compounds merit consideration in this respect as well as in that of their synthetical importance.

As the above formulae show, it was formerly held that the ether used in their preparation played an intrinsic part in the final product. It has, however, now been demonstrated* that reagents of normal reactive properties may be prepared from magnesium and chlorobenzene either without a solvent, or in benzene as solvent. By the addition of dioxane $(O(CH_2CH_2)_2O)$ to ordinary ethereal 'Grignard reagents' Schlenk† was able to prove the existence in the solution of three types of compounds: (*a*) MgR_2, which alone is soluble in dioxane, (*b*) $RMg(hal.)$, (*c*) inorganic $Mg(hal.)_2$, both of the latter being precipitated by dioxane. The efficacy of the 'reagent' is now considered to depend on the equilibrium

$$2RMg(hal.) \rightleftharpoons MgR_2 + Mg(hal.)_2,$$

in which the left-hand member is the chief synthetic agent. By a refinement of his method of precipitation Schlenk arrived at the following results:

Organic halide used	Percentage of RMg(hal.)
CH_3I	87
C_2H_5Cl	25
C_2H_5Br	41
C_2H_5I	43

* Gilman and Brown, *J. Amer. Chem. Soc.* **52**, 3330 (1930).
† W. Schlenk junr. *Ber.* **64**, 734 (1931).

The role of the solvent ether is now seen to be the maintenance of the equilibrium by the retention in solution of all its constituents. The Mg alkyls extracted by Schlenk were solids, and therefore probably of ionic constitution like the alkyls of the alkali metals (p. 51).

(3) *Peroxides and carbides*

The peroxides RO_2 yield H_2O_2 and a normal salt of the metal on treatment with acids. The carbides RC_2 yield acetylene even with water.* X-ray crystal analysis shows that the lattices of both classes of compounds are ionic, containing respectively the ions $O^- — O^-$ and $C^- \equiv C^-$, both of which are colourless (fig. 19). The ionic arrangement is that of a NaCl or MgO lattice distorted in one direction to accommodate the longer diatomic ions. In agreement with this difference the crystals are tetragonal instead of cubic $(c/a \sim 1.2)$.

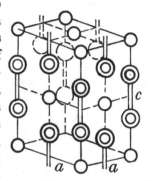

Fig. 19. The tetragonal unit cell of CaC_2 (or BaO_2). ◎, C or O; O, Ca or Ba.

(4) *The valency of mercury*

Mercury is bivalent in both the mercurous and the mercuric series of compounds. That the mercurous ion has the composition $(Hg_2)^{++}$ has been demonstrated as follows:†

(*a*) When mercury is shaken with aqueous silver nitrate some mercurous ions enter the solution, and some metallic silver amalgamates with the mercury. An equilibrium is reached, which may be represented in alternative ways:

$$Hg(liq.) + Ag^+ aq. \rightleftharpoons Hg^+ aq. + Ag(amal.), \qquad (i)$$

$$2Hg(liq.) + 2Ag^+ aq. \rightleftharpoons (Hg_2)^{++} aq. + 2Ag(amal.). \qquad (ii)$$

If the experimental concentrations of Ag^+, mercurous ion and amalgamated silver, in the final equilibrium, are respectively

---

* Beryllium carbide, Be_2C, which yields CH_4 with water, has the anti-fluorite structure (fig. 10) and is probably also ionic (Be^{2+} and C^{4-}).

† Ogg, *Z. physikal. Chem.* **27**, 285 (1898).

a, b and c, then for (i) $a/(bc) = k$, and for (ii) $a/(c\sqrt{b}) = k$. The second relation alone agrees with experiment.

(b) The E.M.F. of the concentration cell*

$$\text{Hg} \mid \tfrac{1}{10} \text{molar 'HgNO}_3\text{'} \mid\mid \tfrac{1}{100} \text{molar 'HgNO}_3\text{'} \mid \text{Hg}$$

is found to be 0·0274 volt at 25°. Since the E.M.F. is given by the general relation $E = \dfrac{0\cdot058}{z} \log \dfrac{c_1}{c_2} = \dfrac{0\cdot058}{z}$ in the above case, it is clear that $z = 2$.

Crystal analysis by X-radiation demonstrates the presence of molecules $\text{Hg}_2(\text{hal.})_2$ in the (tetragonal) crystals of the mercurous halides.†

GROUP III (A)

Boron

(a) The only non-metallic member of this group exhibits tervalency unambiguously in the volatile halides BF_3 (b.p. −101°), and BCl_3 (b.p. 18·2°). BF_3 resembles SiF_4 and PF_5, but differs from CF_4, and SF_6, in being hydrolysed by water. The crystal structure of boric acid B(OH)_3 (p. 229), in which three oxygen atoms are found at equal distances from and co-planar with the boron atom, indicates tervalency also towards oxygen. Electron-diffraction methods applied to BF_3 and BCl_3 show that also in these molecules the halogen atoms are equidistant from and co-planar with the boron atom (p. 47). BH_3 cannot apparently exist stably (see below), but its organic derivatives, such as $\text{B(CH}_3)_3$, are well known. Compounds of tervalent boron show a remarkable power of uniting with various other molecules to yield exceptionally stable addition products:

$\text{BF}_3.\text{NH}_3$ is formed by the direct union of the gases BF_3 and NH_3. It is a colourless solid, which sublimes unchanged, but, like NH_4Cl, undergoes thermal dissociation (v.d. = 23).

$\text{BCl}_3.\text{NH}_3$ closely resembles the fluoride product.

$\text{B(CH}_3)_3.\text{NH}_3$ can readily be distilled without decomposition. Other compounds of this type are $\text{BF}_3.\text{O(C}_2\text{H}_5)_2$‡ and

---

* Ogg, *ibid.* p. 295.　† Havighurst, *J. Amer. Chem. Soc.* 48, 2113 (1926).

‡ On the crystal structure of $(\text{CH}_3)_2\text{O.BF}_3$ and other etherates of BF_3 see Bauer, Finlay and Laubengayer, *J. Amer. Chem. Soc.* 65, 889 (1943); Laubengayer and Finlay, *ibid.* p. 884 (1943).

$BF_3.OC-CH_3$. Compounds of the type $BF_3.ROH$, such as
$/$
OCH_3

$BF_3.CH_3OH$ and $BF_3.C_2H_5OH$ (formed from BF_3 and alcohols) function as strong acids, yielding anions $(BF_3.OR)^-$.

(For an account of the action of BF_3 upon organic compounds and its use as a catalyst, see *Ann. Reports*, 1942, pp. 128 *et seq.* See below for compounds of BH_3.)

(b) *The hydrides of boron.**

	B_2H_6	B_4H_{10}	B_5H_9	B_5H_{11}	B_6H_{10}	$B_{10}H_{14}$
M.p. (° C.)	−165·5	−120	−42·6	−123	−65	99·7
B.p. (° C.)	−92·5	18	48	63	—	213·0

Above are seen the formulae and some physical properties of the best-known boranes. With the exception of decaborane, $B_{10}H_{14}$, all the hydrides are decomposed by water, ultimately yielding boric acid and hydrogen. Although the simplest hydride, diborane, does not come into stable equilibrium with a monomeric species BH_3 on heating, yet in the presence of CO or $N(CH_3)_3$ it yields the compounds BH_3CO (borine carbonyl, b.p. −64°) and BH_3NMe_3 (m.p. 94°) respectively. Diborane reacts with $B(CH_3)_3$, which unlike BH_3 is stable, to give $B_2H_2(CH_3)_4$, but no further substitution can be effected.

Schlesinger and his co-workers discovered a series of metallic borohydrides closely associated in mode of formation and properties with diborane.† When aluminium methyl (prepared from Al and $Hg(CH_3)_2$; m.p. 15°) absorbs diborane at 60° a reaction represented by the following equation readily takes place:

$$Al_2(CH_3)_6 + 4B_2H_6 = 2B(CH_3)_3 + 2Al(BH_4)_3.$$

An entirely analogous reaction between $Be(CH_3)_2$ and B_2H_6 yields the compound $Be(BH_4)_2$. The aluminium borohydride and lithium ethyl (p. 51) react together in benzene solution as follows:
$$Al(BH_4)_3 + 3LiC_2H_5 = 3Li(BH_4) + Al(C_2H_5)_3.$$

* Bell and Eméleus, *Quart. Rev.* II, 132 (1948).

† *J. Amer. Chem. Soc.* **62**, 3421, 3425 and 3429 (1940) and *Ann. Reports*, **38**, 65 (1941).

Alkali-metal borohydrides can also be obtained from their hydrides and diborane:

$$2LiH + B_2H_6 = 2LiBH_4.$$

Al and Be borohydrides boil at 44° and 91° respectively: the Li and Na compounds are crystalline and non-volatile, the latter having the NaCl structure with $(BH_4)^-$ anions, and being recoverable unchanged from its solution in cold water. An examination of Al borohydride by the method of electron-diffraction[*] was considered by its authors to indicate the structure shown in fig. 20:

$$\angle AlBH_a = 180°,$$

$$\angle AlBH_b(H_c \text{ or } H_d) = 85°,$$

$$\angle BAl(BH_4- = 120°,$$

Al, B = 2·14, B, H = 1·27.

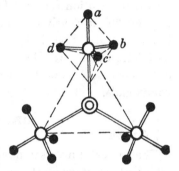

Fig. 20. The structure of $Al(BH_4)_3$. ◉, Al; ○, B; ●, H.

The three B atoms are co-planar with the Al atom, which therefore appears to show a close stereochemical resemblance to boron in BF_3 and other compounds, but the Al, B distance is distinctly longer than would be expected for a normal single bond between these elements. Each boron atom lies at the centre of a slightly distorted trigonal bi-pyramid ($\angle AlBH_b = 85°$) and resembles in this disposition of bonds quinquevalent phosphorus in PF_5.

Owing to the difficulty of fixing the positions of hydrogen atoms by diffraction methods, the correctness of this structure is still open to doubt (see further, p. 236).

* Beach and Bauer, *J. Amer. chem. Soc.* **62**, 3440 (1940).

GROUP III (B)

Thallium

The tendency of valency shown in the more stable compounds to decrease in the B-groups with increasing atomic weight is well illustrated by thallium, the heaviest non-radioactive member of Group III. In the colourless, sparingly chloride TlCl, with crystal structure the same as that of AgCl, the element is clearly univalent. The same valency can be assumed in the soluble hydroxide TlOH and carbonate Tl_2CO_3. Alums with Tl as the tervalent metal have not been prepared, but those with the metal in the univalent position are well known. The trichloride $TlCl_3$, with m.p. 60–70°, is clearly non-ionic; it dissociates on heating to the lower chloride. Like BCl_3 it forms stable addition compounds with NH_3, ethers, etc. The oxide Tl_2O_3, which begins to dissociate only above 800°, is probably the most stable of the tervalent Tl compounds.

GROUP IV (B)

(a) Carbon

The almost universal quadrivalency of carbon exerted indifferently towards all elements combining with it (in particular, to hydrogen, halogens, oxygen and nitrogen) has, since the time of Kekulé, remained a fundamental tenet of organic chemistry. Other non-metallic elements exhibit not only different numerical valencies to hydrogen on the one hand, and halogens on the other (e.g. PH_3, PF_5; SH_2, SF_6), but the no less important characteristic that their compounds with these two types of elements usually differ radically in chemical properties. It is, however, essential to the well-established principle of substitution, first experimentally proved and asserted by Dumas (p. 7), that the interchange of hydrogen and halogen in carbon compounds does *not* occasion a marked change in chemical properties (e.g. CH_4, CCl_4, etc.). This indifference to the nature of its chemical partners is probably not unconnected with the unique capacity of carbon atoms for self-saturation, leading to the existence of homologous series, whose successive members show little if any decline of stability with increasing number of carbon atoms. Stereochemical studies early established the tetrahedral disposition of the four carbon bonds, and modern physical methods have fully confirmed this conception.* Only in rare cases is one of the carbon bonds ionized; examples are found in the carbide ion, $(C\equiv C)^-$, and the negative alkyl ions in certain metallic alkyls, e.g. $CH_3^-.Na^+$ (p. 51). Kekulé's assumption that in the so-called unsaturated carbon compounds the quadrivalency is maintained by the existence of multiple bonds between carbon atoms (e.g. in C_2H_4 and C_2H_2) has been universally accepted by organic chemists almost since his time. The special (additive) reactivity of this type of compound has placed them in a position of unique importance in organic chemistry, and the impression is gained that C, C multiple bonds possess physical distinctions from other multiple bonds. In examining this question the collection of physical data in the following tables is significant.

Table 8. *Bond energies*

(Heats of formation of bonds from atoms—Cal./bond)

	Single	Double	Triple
C, C	83 (C_2H_6)	146 (C_2H_4)	200 (C_2H_2)
C, O	82·6 $((CH_3)_2O)$ 81 (C_2H_5OH)	175 (CH_3COCH_3) 173 (CH_3CHO)	256 (CO-spectroscopic)
C, N	69 (CH_3NH_2) 69 $((CH_3)_2NH)$ 68 $((CH_3)_3N)$	142 (p. 66)	214 (CH_3CN)

It will be noticed that the heat of formation of C, C links increases less rapidly than in proportion to the multiplicity, while the heat for C, O and C, N links increases in approximate proportion to the multiplicity. This relation for C, N links may, however, be fortuitous rather than significant, for in the series N—N, N=N, N≡N the heat of formation increases *more* rapidly than in proportion to the multiplicity and C, N links might be expected to exhibit an intermediate character. In multiple C, C links, less energy being required to break one link

* Brockway and Wall, *J. Amer. Chem. Soc.* **56**, 2373 (1934).

of the bond than to break a single bond, additive reactivity is shown.

Table 9. *Force constants* (dynes/cm. $\times 10^{-5}$)

Linked atoms	Single	Double	Triple
C, C C, O	4·5, C_2H_6 4·5, C_2H_5OH	9·8, C_2H_4 12·3, H_2CO	15·7, C_2H_2 (18·9, CO)
C, N C, S	5·0, $MeNH_2$ 3·0, C_2H_5SH	(12·1) 7·6, CS_2	18·1, HCN —
O, O N, N	3·8, H_2O_2 3·6, N_2H_4	11·7, O_2 —	— 22·8, N_2

Table 10. *Bond-lengths* (Angstrom units)

Linked atoms	Single	Double	Triple
C, C	1·54	1·33	1·20
C, N	1·47	—*	1·15
C, O	1·42	1·20	1·13 (CO)
N, N	1·47	1·24	1·09
O, O	1·48	1·21	—

Percentage contractions	Double	Triple
C, C	13·6	22·0
C, N	—*	21·8
C, O	15·4	20·4
N, N	15·6	25·8
O, O	18·2	—

* See *p. 215*.

It will be recalled that the force constant is numerically a measure of the force required to stretch the linked atoms apart (p. 39). In the light of the above data C, C multiple links are relatively weak. The listed bond-lengths show that the ethylenic linkage is distinctly less compact than would be expected, but the acetylenic linkage is normal.[†]

The calculation of bond energies of carbon compounds[‡]

For a direct computation of bond energies we require to know with accuracy (1) heats of combustion (to CO_2, $H_2O(l)$, N_2, etc.), (2) heats of atomization of the common elements (solid carbon, H_2, O_2, N_2, etc.). By combining the heats of combustion with the

[†] For a relation between bond-length and force constant (Badger's rule) see *J. Chem. Physics*, **2**, 128 (1934) and *ibid.* **3**, 710 (1935).
[‡] Cf. Rossini, *J. Ind. Eng. Chem.* **29**, 1424 (1937).

heats of formation of CO_2 (94·03 Cal.) and liquid water (34·19 Cal. per $\frac{1}{2}H_2$), the heats of formation ΔH_f from the common forms of elements are immediately obtained. Values of ΔH_f used in computing Table 8 are given below in Table 11.

Table 11. *Heats of formation* (from graphite, H_2, O_2, N_2: in Cal.)

(a) C_2H_6 $-20\cdot191$ C_2H_4 $+12\cdot556$
C_3H_8 $-24\cdot75$ C_2H_2 $+54\cdot76$

(b) $(CH_3)_2O$ $-44\cdot20$ CH_3NH_2 $-6\cdot0$ CH_3CN $+19\cdot8$
C_2H_5OH $-56\cdot2$ $(CH_3)_2NH$ $-1\cdot4$ CH_3NCO $-21\cdot23$
CH_3CHO $-45\cdot8$ $(CH_3)_3N$ $+2\cdot3$ C_2H_5NCO $-28\cdot5$
CH_3COCH_3 $-52\cdot5$

(c) CO $-26\cdot394$ H $+51\cdot7$ $-NH_2$ $(\frac{2}{3}NH_3+\frac{1}{6}N)$ $+30\cdot1$
CO_2 $-94\cdot030$ O $+58\cdot7$ $=NH$ $(\frac{1}{3}NH_3+\frac{2}{3}N)$ $+71\cdot3$
H_2O $-57\cdot798$ (g), N $+112\cdot5$ $-OH$ $(\frac{1}{2}H_2O+\frac{1}{2}O)$ $+0\cdot50$
$$$-68\cdot38$ (l) C $+170\cdot9$ $-CH_3$ $+31\cdot20$
NH_3 $-11\cdot040$ $-C_2H_5$ $+24\cdot36$

$$CH_3CH < +68\cdot9$$

Data in (a) and (c) from Thacker, Folkins and Miller, *J. Ind. Eng. Chem.* **33**, 584 (1941); (b) from Kharasch, *Bur. Standards J. Res.* **2**, 359 (1929), and other more recent data.

Under (2) the heats of dissociation of H_2, O_2 and N_2 are accurately known from spectroscopic data (values on p. 113). Although little doubt now persists in the value of the heat of atomization (sublimation) of carbon L_C (see Table 11 (c)), in calculating the bond energies in Table 8 explicit use of L_C has been avoided, but advantage has been taken of the direct determination of the bond energies of C—C and C—H by the method of electron impact and the mass-spectrograph.* These experiments give heats of reaction as follows:

$CH_4 = CH_3 + H$;   101 Cal. $C_2H_6 = 2CH_3$;   82·6 Cal.
$C_2H_6 = C_2H_5 + H$;   96 Cal. $C_4H_{10} = 2C_2H_5$;   77·6 Cal.

The methods of computation employed may be explained by two examples:

(1) C—O.

$C_2H_6 = 2CH_3$; $\Delta H_f = -20\cdot19 + 82\cdot6 = 62\cdot41$.
Total ΔH_f for $2CH_3 + O = 62\cdot41 + 58\cdot7 = 121\cdot11$.
ΔH_f for $(CH_3)_2O$ from heat of combustion $= -44\cdot20$.
Bond energy C—O $= \frac{1}{2}(121\cdot11 + 44\cdot20) = $ **82·6**.

* Stevenson, *J. Chem. Physics*, **10**, 291 (1942).

(2) **C≡N.**

The following heats of combustion (Q) will be employed:

Q (n-butylamine, $C_4H_9NH_2$)   710·6 Cal. (1)

Q (isobutane, $(CH_3)_3CH$)   686·3 Cal. (2)

Q (isobutylidene-n-butylamine,

$C_3H_7CH:N(C_4H_9)$)   1295 Cal. (3)

(1) Kharasch, *Bur. Stand. J. Res.* 2, 359 (1929).

(2) Prosen and Rossini, *J. Res. Nat. Bur. Stand.* 34, 263 (1945).

(3) Coates and Sutton, *J. Chem. Soc.* 1948, p. 1187.

From the sequences

$$C_4H_9NH_2 \rightarrow C_4H_9N\langle + H_2 \rightarrow \text{combustion products,}$$

$$C_3H_7CH_3 \rightarrow C_3H_7CH\langle + H_2 \rightarrow \text{combustion products,}$$

we see that the heat of combustion of each radical is that of the parent compound diminished by the energy released on the removal and burning of 2 hydrogen atoms:

$$Q(C_4H_9N\langle) = 710\cdot6 + 2(\text{N—H}) - 103\cdot4 - 68\cdot4,$$

$$Q(C_3H_7CH\langle) = 686\cdot3 + 2(\text{C—H}) - 103\cdot4 - 68\cdot4,$$

whence
$$Q \text{ (radicals)} = 1053\cdot3 + 2(\text{N—H}) + 2(\text{C—H}).$$

Now
$$Q \text{ (radicals)} - Q \text{ (combined radicals)} = (\text{C≡N}).$$

Therefore

$$(\text{C≡N}) = 1053\cdot3 - 1295 + 2(\text{N—H}) + 2(\text{C—H}).$$

Taking (C—H) as 99 Cal. and (N—H) = 93 Cal. = $\frac{1}{3}$ (heat of formation of NH_3 from atoms), we find (C≡N) = **142·3 Cal.**

(On the heat of atomization of carbon, L_C, and the heat of dissociation of N_2 see Douglas, *J. Physical Chem.* 59, 109 (1955), and *Ann. Reports*, LIV, 1957, p. 81.

The free energy of bonds

Undoubtedly the standard *free* energy of bonds $F°$ would provide a more satisfactory basis on which to estimate their strength and stability than the standard *total* energy at room temperature $E°_{298}$ usually solely employed, which in fact conceals the influence of temperature on strength of linking. Free energy F and total energy E are related by the fundamental equations

$$F° = E° - TS° = E° + T\frac{dF°}{dT} \quad (S° = \text{standard entropy}). \quad \text{At abso-}$$

lute zero $F_0^\circ = E_0^\circ$, and since E° does not change very rapidly with temperature, $E_{298}^\circ \sim E_0^\circ$; the use of E_{298}° as a measure of F_0° is to this extent justified. There is, however, considerable difference between E_{298}° and F_{298}°, as seen for example in the hydrogenation of ethylene (Table 14 and p. 200):

$$C_2H_4 + H_2 = C_2H_6, \quad \Delta F_{298}^\circ = -24{\cdot}1 \text{ Cal.,}$$
$$\Delta E_{298}^\circ = -32{\cdot}6 \text{ Cal.,}$$

and this difference must usually increase rapidly as temperature rises, for the product TS° changes more rapidly than E°. Unfortunately the available data on free energies of atomization, especially of solid elements (e.g. carbon and sulphur), are at present inadequate for a general survey of the free energy of bonds, but the following examples of the free energy of formation (from graphite and H_2) and of hydrogenation (by H_2) of simple carbon compounds are of interest in respect to C, C and C—H bonds. The data are drawn from the valuable calculations of Kassel (*J. Amer. Chem. Soc.* **55**, 1351 (1933)), Thomas, Egloff and Morrell (*J. Ind. Eng. Chem.* **29**, 1260 (1937)) and Thacker, Folkins and Miller (*J. Ind. Eng. Chem.* **33**, 584 (1941)).[*]

Table 12. *Free energies of formation ΔF° (Cal.)*
from graphite and H_2

° K.	CH_4	C_2H_6	C_3H_8	C_2H_4	C_2H_2
300	-12·1	-7·8	-5·6	16·3	49·9
500	-7·9	1·25	8·3	19·3	47·0
1000	4·5	26·25	46·1	28·4	40·5
1200	9·8	36·4	—	32·1	37·9
1500	—	—	—	35·1	34·1

Table 13. *Relative stability*

	300° K.	500° K.	1000° K.	1200° K.	1500° K.
Stability ↑ increasing	CH_4 C_2H_6 C_2H_4 C_2H_2	CH_4 C_2H_6 C_2H_4 C_2H_2	CH_4 C_2H_6, C_2H_4 C_2H_2 —	CH_4 C_2H_4 C_2H_6, C_2H_2 —	(CH_4) C_2H_4, C_2H_2 (C_2H_6) —

[*] The last paper contains reliable data on the free energies of formation of CO, CO_2, CS_2, H_2O, H_2S, NH_3, SO_2 and SO_3, as well as of paraffins CH_4 to $C_{10}H_{22}$, and olefines C_2H_4 to $C_{10}H_{20}$.

Tables 12 and 13 show that with rise of temperature the C, C multiple bond becomes more stable relatively to the C, H link. In the temperature range 300° to the neighbourhood of 1000° olefines will tend to polymerize; above about 1000° unsaturates become increasingly stable, and dehydrogenation of paraffins and olefines sets in. CH_4 evidently retains a pre-eminent stability throughout the temperature range 300–1500° K.

Table 14. *Free energy of hydrogenation ΔF° (Cal.)*

System	300° K.	500° K.	1000° K.	1200° K.	1500° K.
$C_2H_2 + H_2 = C_2H_4$	– 33·6	– 27·7	– 12·1	– 5·8	1·0
$C_2H_4 + H_2 = C_2H_6$	– 24·1	– 18·0	– 2·15	4·3	—
$C_2H_6 + H_2 = 2CH_4$	– 16·4	– 17·0	– 17·2	– 16·8	—
$C_3H_8 + H_2 = CH_4 + C_2H_6$	– 14·3	– 14·0	– 15·4	—	—

All the reactions of Table 14 essentially exemplify the change

$$C—C + H_2 \rightarrow 2CH.$$

It is again evident from this Table that C—C in a *multiple* C, C link increases greatly in strength relatively to C—H with rise of temperature. On the contrary C—C in the paraffins undergoes practically no change in strength relatively to C—H in the temperature range indicated. (For values of ΔE° in hydrogenation see Table 32, p. 200.)

Although in almost all cases where necessary the assumption of multiple links not only maintains the quadrivalency of carbon but yields bond-diagrams in agreement with experimental facts, there remain a few compounds where quadrivalency cannot apparently be retained without violating chemical or stereochemical principles. Williamson (p. 11) was prepared to accept quadrivalent oxygen in carbon monoxide, but it is now known that around 4-bonded atoms of both carbon and oxygen the bonds are disposed tetrahedrally (cf. pp. 40, 41, 63). In the bond-diagram C≡O it would clearly be impossible to preserve such a disposition around either atom. If, on the other hand, the oxygen retains its more normal bivalency, then carbon is also bivalent in carbon monoxide. A similar stereochemical difficulty arises in assigning formulae to the isonitriles RNC and the fulminates RONC.

For this reason it was until recently commonly assumed that these types of compound and CO contained bivalent carbon (*p. 114*). In view of the indifference of carbon to the nature of the atoms with which it unites (see above) it would be contrary to all expectation that CH_2 and $C(hal.)_2$ should resist all attempts at isolation while CO is a very stable molecule. Much light is thrown on these long-debated problems by modern physical methods. C, O in carbon monoxide is 1·13; its heat of formation from atoms is 256 Cal./mol.;[*] its force constant is 18·9 units. Consultation of Tables 8, 9, 10 above shows that all these values support the diagram C≡O. C, N in the isonitriles is found to equal exactly C, N in the nitriles RCN (1·16),[†] which suggests the diagram R—N≡C. It will be seen that these more recent formulae appear only to substitute grave new difficulties for the old; it is one of the triumphs of the new electronic theory of valency to have explained their rectitude (*p. 123*).

The oxide CO_2, the sulphide CS_2 and the oxysulphide COS all have the linear arrangement to be expected from the application of the tetrahedral model to the structures X=C=X. In agreement with their symmetrical shape CO_2 and CS_2 have zero electric moment. Phosgene, $COCl_2$, is triangular, $O{=}C{<}^{Cl}_{Cl}$, with ∠OCCl 125°,[‡] also agreeing with tetrahedral theory.

(b) *Silicon*

No stable compounds are known in which silicon exerts a valency less than 4. The hydrides (silanes) SiH_4, Si_2H_6, Si_3H_8 and Si_4H_{10} are analogous in formulae and molecular structure to the lower paraffins, but all are inflammable on contact with air, and explode on mixing with chlorine. The halides (e.g. SiF_4, $SiCl_4$)

---

[*] Kynch and Penney, *Proc. Roy. Soc.* A, **179**, 214 (1941).
[†] Brockway, *J. Amer. Chem. Soc.* **58**, 2516 (1936).
[‡] Robinson, *J. Chem. Physics.* **21**, 1741 (1953).

are rapidly hydrolysed by water. The sulphide SiS_2* forms fibre macromolecules, probably of the units†

and is therefore intermediate in chemical and crystalline structure between crystalline CO_2, which is composed of individual CO_2 molecules, and the three-dimensional macromolecular SiO_2 described on p. 40.

(c) Tin and lead

Both metals show a complete series of salts derived from the basic oxides SnO and PbO, and in exhibiting such a well-marked bivalency they illustrate the rule that valency tends to diminish in the B-groups with increasing atomic weight. Tin still shows quadrivalency (sometimes as an ion) as its most stable state, whereas very few stable compounds are known of quadrivalent lead.

The gaseous hydrides SnH_4 and PbH_4 have been obtained in small quantities, and prove to be very unstable (Paneth).‡ It may here be noted that, boron and gallium being excepted, elements forming *volatile* hydrides are contained in the B families of Groups IV to VII inclusive. The oxides SnO_2 and PbO_2 both form ionic lattices (R^{4+} and $O^{=}$) with the rutile (TiO_2) structure (fig. 21). It is of great interest that GeO_2, the oxide of the element (Mendeleeff's eka-silicon) between Si and Sn, is dimorphic, one form being isomorphic with quartz, and the other (with ionic lattice) isomorphic with SnO_2.§ PbO_2 and Pb_3O_4‖ are stable compounds, with 4-valent Pb, and probably the only examples of the plumbic ion Pb^{4+}. The tetra-alkyls of Sn and Pb are stable liquids, prepared by the action of Grignard reagents or other alkyls on the metallic halides. In this type of reaction Pb tetra-

---

* Tiede and Thimann, *Ber.* 59, 1703 (1926).

† Zintl and Loosen, *Z. physikal. Chem.* A, **174**, 301 (1935); also *Ann. Reports*, 1935, p. 209; cf. SeO_2, p. 87.

‡ PbH_4: Paneth and Norring, *Ber.* **53**, B, 1693 (1920); SnH_4: Paneth, *Ber*, **57**, B, 1877 (1924).

§ *Ann. Reports*, 1934, p. 119.

‖ Gross, *J. Amer. Chem. Soc.* **65**, 1107 (1943).

alkyls are obtained from the *lower* chloride $PbCl_2$ by the remarkable disproportionation

$$2PbCl_2 + 4RMgBr = PbR_4 + 2Pb + 2MgCl_2 + 2MgBr_2.$$

The tetrahedral disposition of the Pb—C bonds in these alkyls has been proved by the use of electron-diffraction methods.*

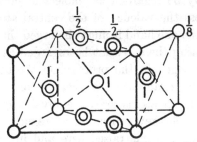

Fig. 21. The tetragonal unit cell of rutile (TiO_2). ◎, O; ○, Ti.

$SnCl_4$, a liquid (b.p. 114°), is clearly non-ionic and thus is analogous with CCl_4 and $SiCl_4$, but treatment with a limited amount of water produces the crystalline hydrate $SnCl_4.5H_2O$, which very probably contains the (hydrated) ion Sn^{4+}. Hydrolysis proceeds in excess of water with the ultimate formation of $Sn(OH)_4$. Similar behaviour is found with stannous chloride (b.p. 600°), which, like the (insoluble) dihalides of lead, forms a non-ionic lattice when anhydrous, but also yields the hydrate $SnCl_2.2H_2O$ containing the stannous ion Sn^{++} (cf. p. 118). It is to be noted that while the non-ionic dihalides of Group II, e.g. $BeCl_2$, Hg_2Cl_2, $HgCl_2$, are all linear molecules, SnX_2 and PbX_2 are triangular; $\angle BrPbBr$ in $NH_4Br.2PbBr_2$ has been measured as about 86°.† The following crystalline tetrahalides, belonging to the cubic system, all exhibit regular tetrahedral structure, as proved by X-ray analysis: SiI_4, SnI_4; and from the A-group, $TiBr_4$, TiI_4, $ZrCl_4$ (and CeF_4).

* Brockway and Jenkins, *J. Amer. Chem. Soc.* 58, 2036 (1936).

† Powell and Tasker, *J. Chem. Soc.* 1937, p. 119.

Complex halides of Si, Sn and Pb

The tetrahalides of these elements all show a marked tendency
to unite with simple halide ions in the reaction

$$RX_4 + 2X^- \rightarrow RX_6^=;$$

as examples we have $SiF_6^=$, $SnCl_6^=$, $PbCl_6^=$, etc. As in most complex
ions (which may be regarded as 'molecular compounds') it is
difficult to assess the valency of the central atom on classical
principles. On the one hand it might be urged that, as the above
reaction generating them cannot be regarded as oxidative, the
valency is unchanged in the process; on the other hand it is
known from X-ray examination of crystals containing these ions
that the six halogen atoms are all equivalently united to the
central atom, which they surround in a regular octahedron (cf.
SF_6, p. 47). The problems raised here and in numerous other
similar cases receive an acceptable solution only by applying the
modern electronic theory (*pp. 152, 168*).

GROUP IV (A)

(a) *Ti, Zr and Th*

As is to be expected in members of the A-group, quadrivalency
is maintained in all the more stable compounds of these metals.
The compounds of thorium are almost all ionic, and contain the
ion Th^{4+}. ThO_2 appears to be fully basic in function, and is the
only dioxide that can be so described. Ti exhibits a well-marked
tervalency, while very unstable tervalent salts of Zr are known.
$TiCl_3$ and $Ti_2(SO_4)_3$ are isomorphic with the corresponding ferric
salts; both are extensively used in volumetric analysis as powerful reducing agents. The titanous ion is coloured violet in solution
(pp. 26,*158*). TiF_4 (b.p. 284°), $TiCl_4$ (b.p. 135°) and TiO_2 very closely
resemble the corresponding compounds of Sn. The lower valencies
of Ti and Zr are attributable to the transitional character of
these elements, in the wider sense in which that term is now
understood (*p. 158*).

(b) *The position of cerium*

Owing to its quadrivalency in compounds derived from its
principal oxide CeO_2, Mendeleeff placed this metal in Group IV,

between Zr and Th. In the light of greater knowledge of the chemistry of these elements this position is not satisfactory, for the small number of ceric salts known all show powerful oxidizing properties, and the (hydrated) ceric ion Ce^{4+} is yellow. The best known salt $Ce(SO_4)_2$ is extensively used in volumetric analysis as an oxidizing agent of power almost equal to that of permanganate (the potential of Ce^{4+}/Ce^{3+} is -1.44 volt for molar solution, and of $(MnO_4^- + 8H^+)/Mn^{++}$, -1.48 volt). The cerous salts, of which a large number are known, and which contain the ion Ce^{3+}, closely resemble the compounds of the neighbouring rare-earth elements, of which group Ce is to be regarded as the first member. Ce resembles Mn in the point that while the stable salts are derived from the lower oxide Ce_2O_3, this oxide itself is readily oxidized at low temperature by atmospheric oxygen to CeO_2. The following gradation of properties in the rare-earth elements next to cerium is noteworthy:

Ce: Ce_2O_3 readily oxidized to the more stable CeO_2, which gives rise to a few salts.

Pr: Pr_2O_3 stable to oxidation; PrO_2 has the character of a peroxide; and forms no salts.

Nd: NdO_2 so unstable that its existence has been doubted.

It is clear that the quadrivalency dies out gradually.

GROUP V (B)

(1) *The elementary molecules N_2, P_4 and As_4; oxides of P and As*

The series of bond-lengths N—N $= 1.47$ A. in hydrazine, $NH_2.NH_2$, N$=$N $= 1.20$ A. in azobenzene, and 1.09 A. in N_2, proves that in the last molecule we must assume a triple link. Liquid yellow phosphorus and yellow arsenic are built of molecular units R_4, the four constituent atoms being placed at the corners of a regular tetrahedron, so that the bond-angles are only $60°$, and each atom tervalent (fig. 7, p. 42).* It is interesting to contrast the chemical activity of singly-bonded phosphorus with the inertness of triply-bonded nitrogen. The vapour of red phos-

* *Ann. Reports*, 1939, p. 164.

phorus below 100° contains the molecule P_2,* in which P, P is
1·895.† In view of the contractions listed in Table 10, p. 64, the
contraction in this case of 14·2 per cent from the single bond in
P_4 (2·21) might suggest a double bond P=P as in C_2, rather than
a triple bond as in N_2. It may, however, be noted that the contraction between S—S (2·08 in S_8) and S=S very probably
present in S_2 (S, S, 1·88) is only 9·7 per cent.‡ The structures of
the oxides of P and As (figs. 22, 23) are derived from the framework of the elementary molecules.§ In R_4O_6 the six direct bonds
R—R along the sides of the tetrahedron of R_4 are replaced by

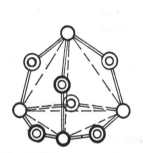

Fig. 22. The structure of P_4O_6.
◎, O; ○, P.

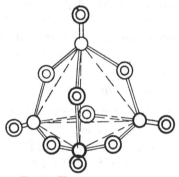

Fig. 23. The structure of P_4O_{10}.
◎, O; ○, P.

bent oxygen 'bridges' R—O—R, in which P—O = 1·62 A. In
R_4O_{10} the additional four oxygen atoms are directly attached,
by multiple bonds (P, O = 1·39), to each R atom. Tervalency
is thus demonstrated in the oxides R_4O_6 and a higher valency
in the oxides R_4O_{10}. It may also be seen that by assuming fewer
oxygen bridges or directly-bonded oxygen atoms we can, without
changes of valency, explain the structure of other oxides, such
as P_4O_8, the separate existence of which has been proved.||

    * Melville and Gray, *Trans. Faraday Soc.* **32**, 271 and 1026 (1936).
    † Herzberg, *Canadian J. Res.* **18**, A, 139; *Amer. Abstr.* **34**, 7178 (1940).
    ‡ Olsson, *Z. Physik*, **100**, 656 (1936). On the structures of red and
black phosphorus see Hultgren, Gingrich and Warren, *J. Chem. Physics*,
3, 351 (1935).
    § Hampson and Stosick, *J. Amer. Chem. Soc.* **60**, 1814 (1938).
    || Miller, *Proc Roy. Soc. Edin.* **46**, 239 (1925).

(2) *The valency of phosphorus and arsenic in other compounds*

In the very stable gases PF_5 (b.p. $-75°$) and AsF_5 (b.p. $-50°$) we have clear proof of quinquevalency, for electron-diffraction methods show that all the R—F bonds are equal in length (fig. 13). The same valency appears to be shown in POF_3 and $POCl_3$, both of which are tetrahedral in shape, and the P-(hal.) bonds all equal in length.* The complex crystal structure of PCl_5 has already been mentioned (p. 48). The valency of P may well be 5 in both the ionic units PCl_4^+ and PCl_6^-, as it must be assumed to be in the phosphonium ion PH_4^+. Crystalline PBr_5 appears to be built of the ions $(PBr_4)^+$ and Br^-.† It appears doubtful whether $AsCl_5$ exists.

We may here summarize the results of the electron-diffraction investigations upon typical compounds of phosphorus, containing P, O and P, S bonds (cf. Stosick, *J. Amer. Chem. Soc.* **61**, 1132 (1939)):

	P, O′	P, O″ or P, S		
$P_4O_6'O_4''(P_4O_{10})$	1·62‡	1·39‡	∠O′PO″	101·5° ± 1°
			∠PO′P	123·5° ± 1°
$P_4O_6S_4$	1·61	1·85	∠OPS	116·5° ± 1°

‡ The bond-lengths are confirmed in *crystalline* P_4O_{10} by de Decker, *Rec. trav. chim.* **60**, 153 and 413 (1941).

The compound $P_4O_6S_4$, phosphorus oxysulphide (colourless, m.p. 205°), is prepared by heating together P_4O_6 and the equivalent weight of sulphur§ just as P_4O_{10} is obtained by treating P_4O_6 with oxygen:

	P, O		P, S
POF_3	1·56	PSF_3	1·85
$POCl_3$	1·58	$PSCl_3$	1·94
$POFCl_2$	1·54		
POF_2Cl	1·55		

From Table 6 we find (P—O) = 1·84, and (P—S) = 2·14. It is

* For ED investigation of $POCl_3$, POF_2Cl, $POFCl_2$, POF_3, PF_3Cl_2 and PF_5 see Brockway and Beach, *J. Amer. Chem. Soc.* **60**, 1836 (1938).

† Powell and Clark, *Nature*, **145**, 971 (1940).

§ Thorpe and Tutton, *J. Chem. Soc.* **59**, 1023 (1891).

thus certain that all the compounds listed above contain *multiple* P, O or P, S bonds. A contraction of (P—O) and (P—S) by 15 per cent gives 1·56 and 1·82 respectively: the data of Table 10 suggest that these will be the expected values of (P=O) and (P=S). The bond-diagrams of the oxy-halides and thio-halides are thus O=PX$_3$ and S=PX$_3$, with 5-valent phosphorus. P, O" in P$_4$O$_{10}$ obviously stands apart, the distance 1·39 being 24 per cent less than (P—O). Such a large contraction indicates a *triple* bond P≡O (Table 10), but it appears impossible to incorporate this in the known structure of the oxide (fig. 23). On the other hand, the P, S distance in P$_4$O$_6$S$_4$ is close to that expected for P=S.

Towards hydrogen a maximum valency of 3 is shown in the hydrides PH$_3$ and AsH$_3$. Phosphorus and arsenic thus afford examples of the 'rule of eight' (p. 29). By treating POCl$_3$ with (a) NH$_3$, (b) the Grignard reagent CH$_3$MgBr, the interesting compounds OPN and the phosphine oxide (CH$_3$)$_3$PO can be obtained. In these compounds phosphorus appears to exert its higher valency.

(3) *Nitrogen*

(a) Nitrogen is clearly tervalent in NH$_3$, and appropriate structural formulae assuming this valency can be written for the following simple and stable compounds: NH$_2$OH, NH$_2$.NH$_2$, NOF, (NOCl); nitrides such as Li$_3$N and Mg$_3$N$_2$; (CN)$_2$ and the organic nitroso-compounds; the ion NO$_2^-$. The gaseous NF$_3$ (b.p. −120°) proves to be unreactive, like CF$_4$. It is insoluble in and unattacked by water, or aqueous potassium hydroxide.[*] No higher fluoride is known (cf. PF$_5$).

(b) *The oxides of nitrogen*

N$_2$O. From spectral observations the structure is certainly NNO, and the molecule is linear (p. 47).[†] These physical results appear to demand the bond-diagram N≡N=O. Such a diagram is also indicated by a rational interpretation of the chemical methods of preparing nitrous oxide.

---

[*] Ruff, Fischer and Luft, *Z. anorg. Chem.* **172**, 417 (1928).

[†] For examination of N$_2$O by electron-diffraction see Schomaker and Spurr, *J. Amer. Chem. Soc.* **64**, 1184 (1942).

(i) By the classical method from NH_4NO_3. Since the normal dissociation of NH_4 salts precedes decomposition, the reaction is essentially between anhydrous, i.e. un-ionized, HNO_3 and NH_3, thus:

$$O=N\overset{O}{\underset{OH}{\diagup}} + \overset{H}{H\text{--}}N\overset{}{\diagup} = O=N\equiv N.$$

(ii) From nitrous acid and hydroxylamine. This reaction proceeds smoothly in acid solution at room temperature, and can be represented thus:

$$N\overset{O}{\underset{OH}{\diagup}} + \overset{H}{H\text{--}}N^+\text{--}OH...X^- = N\equiv N=O + HX.$$

(iii) By the oxidation of hydroxylamine in alkaline solution, e.g. by ammoniacal cupric solutions. The reaction undoubtedly gives rise to hyponitrous acid as an intermediate product, thus:

$$2\left[\overset{H}{\underset{H}{\diagup}}N\text{--}OH\right] + O_2 = 2\left[\overset{OH}{\diagdown}\underset{H}{\overset{N}{|}}\overset{OH}{\diagup}\right] = \begin{matrix}N\text{--}OH\\ \| \\ OH\text{--}N\end{matrix} + 2H_2O.$$

Hyponitrous acid exists only in the *trans*-form shown above:*

$$\begin{matrix}N\text{--}OH\\ \| \\ OH\text{--}N\end{matrix} \rightarrow \left[N\overset{OH}{\underset{|}{\diagdown}}\underset{N}{\underset{\|}{|}}\overset{}{\diagdown H}\right] \rightarrow \begin{matrix}N\\ \| \\ N\\ \| \\ O\end{matrix} + H_2O.$$

The spontaneous transformation of the hypothetical dihydroxyethylene and vinyl alcohol are seen to be closely parallel to the above changes in hyponitrous acid, except that valency restrictions preclude the loss of water from the aldehydes:

$$\left[\begin{matrix}HC\text{--}OH\\ \| \\ OH\text{--}CH\end{matrix}\right] \rightarrow \begin{matrix}CH_2OH\\ | \\ CH\\ \| \\ O\end{matrix} \quad ; \quad \left[\begin{matrix}CH_2\\ \| \\ CHOH\end{matrix}\right] \rightarrow \begin{matrix}CH_3\\ | \\ CH\\ \| \\ O\end{matrix}$$

NO. This compound, which in its physical properties and stability closely resembles N_2 and O_2, and apart from its characteristic addition of oxygen shows little chemical reactivity, long defied decisive chemical interpretation. It could only be said that

* Hunter and Partington, *J. Chem. Soc.* 1933, p. 309.

it appeared to demand the bivalency of nitrogen (for modern views; see *p. 127*).

$NO_2(N_2O_4)$. The constitution of NO_2 is as puzzling as that of NO. Physical methods prove it to be triangular in form, and the structure $O{=}N{=}O$, which is formally analogous to that of CO_2, is improbable. Unlike NO the compound is deeply coloured, and readily polymerizes: $2NO_2{\rightleftharpoons}N_2O_4$. X-ray analysis of the colourless, solid N_2O_4* shows the structure $O_2N.NO_2$, for which we may write the diagram

$$O{\gt}N{-}N{\lt}O \text{ (with O's)}$$

The weakness of the N—N bond is indicated by the small free energy change of 1·20 Cal./mol. on polymerization. The N—N bond is also abnormally long (1·6–1·7). N_2O_4 dissolves in solutions of alkalis to give equivalent amounts of nitrite and nitrate ions thus:

$$\text{(diagram)} \quad +O^- = \left[\text{nitrate}\right]^- \quad \left[\text{nitrite}\right]^-$$

N_2O_3. In a gaseous mixture of NO and NO_2 in equimolecular proportions it is found that the equilibrium $NO + NO_2 \rightleftharpoons N_2O_3$ is established, and at 25° and 1 atm. total pressure there is 10·5 per cent of N_2O_3.[†] The union is accompanied by the liberation of only 0·44 Cal./mol. free energy. On cooling the gaseous mixture, both NO_2 and N_2O_3 liquefy, but not NO, so that the blue-green liquid formed contains excess of liquid NO_2. It is therefore uncertain whether the colour is due to N_2O_3 alone or to its solution in NO_2. However, the colour persists on solidification. Of the two alternative diagrams $O{=}N{-}O{-}N{=}O$, $O{\gt}N{-}N{=}O$, the second is preferred, since alkyl nitrites $R{-}O{-}N{=}O$ are almost colourless.[‡] The weakness of the N—N bond is again indicated by the low free energy of the reaction producing it.

* Hendricks, *Z. Physik*, **70**, 699 (1931).
† Abel and Proisl, *Z. Elektrochem.* **35**, 712 (1929).
‡ Wieland, *Ber.* **54**, 1781 (1921).

N_2O_5. This colourless solid oxide is prepared by vigorous dehydration of nitric acid, and is the anhydride of that acid. Even at room temperature it decomposes as follows:

$$N_2O_5 \rightarrow N_2O_3 + O_2 \rightarrow N_2O_4 + \tfrac{1}{2}O_2.$$

Its crystal structure is found by X-ray diffraction to consist of equal numbers of two discrete units: one is identified as NO_3^- by its plane triangular form, and the other, which is linear with a nitrogen atom midway between two oxygen atoms, as the *nitronium* cation NO_2^+. This result does not overstrain classical theory, for we may freely concede analogies between the tetrahedral CH_4 and NH_4^+ on the one hand, and between the linear $O{=}C{=}O$ and $(O{=}N{=}O)^+$ on the other. The salt $(NO_2)^+ . HSO_4^-$ is an important constituent of a mixture of concentrated sulphuric and nitric acids. It should be noted that the structures of *gaseous* N_2O_4 and N_2O_5 remain undecided.

Omitting N_2O, the oxides are closely connected by spontaneous chemical reactions, as the following scheme indicates:

$$
\begin{array}{c}
\qquad\qquad\qquad -O_2 \\
\qquad\qquad\;\downarrow\;-NO \\
NO \quad N_2O_4 \longrightarrow NO_2 \longleftarrow N_2O_3 \qquad N_2O_5 \\
\quad\underline{\qquad O_2 \qquad}\uparrow\; +O_2
\end{array}
$$

It appears from this that the most stable valency-state of nitrogen in these oxides, at room temperature, is not the quinquevalent state, present in all oxides with N_2 in their formulae, but the unknown state in NO_2. At elevated temperatures NO_2 gives place to NO as the most stable oxide. The colourless crystalline substance, long known as 'lead chamber crystals', and originally formulated as nitrosulphonic acid NO_2HSO_3, has been proved to be a salt, $(NO)^+ . HSO_4^-$. Numerous other salts of the oxycation $(NO)^+$ have been isolated, and it has been shown that the crystal structure of $(NO).ClO_4$ closely resembles that of ammonium perchlorate, NH_4ClO_4.*

(4) *Hydrazoic acid and the azides*

Formerly the acid was formulated as $H{-}N{\bigwedge_{N}^{N}}$, but it was

* Klingenberg, *Rec. trav. chim.* **56**, 749 (1937).

shown* that in its reactions the acid resembles nitric acid. In attacking Zn or Fe no hydrogen is liberated, and the products of the interaction include, besides the metallic azides, reduction products such as N_2, NH_3 and N_2H_4. A very limited amount of hydrogen is liberated in reaction with Mg (cf. HNO_3). The acid attacks copper; and a mixture of the acid with concentrated HCl dissolves gold, producing auric chloride. Ferrous azide is oxidized to ferric azide by heating with the acid. In comparison with $HO-N^V{\Large\langle}_O^O$ we write $HN=N^V\equiv N$. This view of the constitution of the azides was confirmed by X-ray examination, which showed that the N—N—N group is linear. In the very interesting substance cyanuric triazide, $C_3N_3(N_3)_3$ (see also *p. 215*),

a similar linear structure was discovered for the azide group.†
Examination by electron-diffraction of HN_3 further confirms these results.‡

GROUP VI (B)

(a) Oxygen

In such simple compounds as H_2O, CH_3OH, $(CH_3)_2O$, F_2O and Cl_2O, oxygen exhibits clearly the bivalency which is maintained in a large number of its known compounds. In these compounds of bivalent oxygen the 'valency angle', i.e. the angle between the two bonds, is found to lie between 105° and 110°, close to the 'tetrahedral' angle 109°, and gives to the above molecules a triangular outline.

Oxonium compounds. Friedel in 1875 showed that ethers, e.g. $(CH_3)_2O$, could form unstable salts with acids (e.g. HCl) and with

---

* Franklin, *J. Amer. Chem. Soc.* 56, 568 (1934).

† Knaggs, *Nature*, 134, 138 (1934) and *Proc. Roy. Soc.* A, 150, 576 (1935).

‡ Schomaker and Spurr, *J. Amer. Chem. Soc.* 64, 1184 (1942).

CH_3I, of the form $[(CH_3)_2OX]^+ . Y^-$. Later, in 1899, Collie and Tickle were able to prepare relatively much more stable salts from dimethyl-γ-pyrone, such as the hydrochloride (m.p. 152°), which is given the formula (see also *p. 217*).

$$\left[HO-C \begin{array}{c} CH-C(CH_3) \\ \diagdown O \\ CH=C(CH_3) \end{array} \right]^+ \ Cl^-$$

$K_b = 3 \cdot 0 \times 10^{-14}$ (cf. urea, $K = 1 \cdot 5 \times 10^{-14}$).

It is now generally recognized that the proton, whose energy of formation from H_2 may readily be computed to be 360 Cal., is far too unstable to exist free in contact with water, with which it undoubtedly unites to yield the 'hydrogen ion' H_3O^+. The reactions

$$H_2O + H^+ \rightarrow H_3O^+ \quad \text{and} \quad H_3N + H^+ \rightarrow H_4N^+$$

are seen to be completely analogous, and the parent ion H_3O^+ has received the name 'oxonium'. That the non-ionized group in the oxonium ion is of pyramidal shape (with apical oxygen) is rendered almost certain by analogy with the more stable sulphonium salts (see below). X-ray analysis of oxides such as BeO or ZnO with the wurtzite or zinc-blende structure (p. 41), and of basic beryllium acetate (p. 56), proves that in these non-ionic compounds each oxygen atom is *tetrahedrally* linked to four metal atoms. The Be, O distances in BeO and in the basic acetate are found to be identical (1·65 A.).

Ozone and hydrogen peroxide. Since these compounds could be held to react chemically as 'oxidized oxygen' and 'oxidized water' respectively, it was frequently urged that their structural formulae should not show all the oxygen atoms as equivalently bonded. Mainly for such reasons formulae such as O=O=O (I) and $\begin{array}{c} H \\ \diagdown \\ H \end{array} O=O$ (I) were often preferred to $\begin{array}{c} O-O \\ \diagdown / \end{array}$ (II) and HO—OH (II), in which the oxygen retains bivalency. Since perdisulphuric acid ($H_2S_2O_8$) may be synthesized by the smooth interaction of chlorosulphonic acid ($Cl.SO_3H$) and anhydrous H_2O_2* this (dibasic) acid must be regarded as the disulphonic

* d'Ans and Friederich, *Ber.* **43**, 1880 (1910); *Z. anorg. Chem.* **73**, 325 (1912).

82   VALENCY PROBLEMS: II

derivative of H_2O_2, i.e. as $O_2(SO_3H)_2$. X-ray analysis of potassium persulphate shows the structure O_3S—O—O—SO_3 for the S_2O_8 ion, with O, O = 1·46 and $\angle SOO$ = 122°. We have here firm support for the formula II for *anhydrous* H_2O_2, which is also indicated by computation of the electric moment.* A detailed examination of the infra-red absorption spectrum of anhydrous H_2O_2 appears to rule out a planar molecule, and hence the form I. The O, O distance is found to be close to 1·48 (cf. $S_2O_8^=$ above) and the skewed (dihedral) configuration $^H\diagdown_{O-O}\diagup^H$ assumed by Sutherland and Penney is confirmed.† This configuration is placed beyond doubt by the recent X-ray examination of 'hyperol', the compound of H_2O_2 with urea.‡ $\angle OOH$ is found to be 101·5°, the dihedral angle 106°, and O, O 1·46. In *aqueous solution* the feeble ionization to the anion $(HO_2)^-$ may well provide the means for the establishment of a tautomeric equilibrium:

$$(OH)_2 \rightleftharpoons H^+ + HO_2^- \rightleftharpoons H_2O:O$$

(cf. *p. 127*).

Apart from the first consideration above (which is by no means conclusive) little *chemical* objection can be taken to the structure II (an equilateral triangle of singly-bonded atoms) for ozone, which is certainly not more stable than cyclopropane $\begin{smallmatrix} CH_2-CH_2 \\ \diagdown \ \diagup \\ CH_2 \end{smallmatrix}$ wherein atoms normally 'tetrahedral' submit to valency angles of 60°; this angle is also found in elementary phosphorus, P_4 (p. 42). If this equilateral structure were correct it is certain that the infra-red (absorption) spectra of ozone and cyclopropane would be closely related, especially as CH_2 and O differ little in mass. The two spectra are, however, found to have nothing in common; the spectrum of ozone, on the other hand, does much resemble those of NO_2 and NOCl,§ and detailed study indicates

* Sutherland and Penney, *Trans. Faraday Soc.* **30**, 898 (1934); *J. Chem. Physics*, **2**, 492 (1934).

† Zumwalt and Giguère, *J. Chem. Physics*, **9**, 458 (1941).

‡ Hughes, Giguère and Lu, *J. Amer. Chem. Soc.* **63**, 1507 (1941).

§ Infra-red, Sutherland and Penney, *Proc. Roy. Soc.* A, **156**, 654 and 678 (1936); ultra-violet, Price and Simpson, *Trans. Faraday Soc.* **37**, 106 (1941).

that $\angle OOO$ is not far from $125°$. Electron-diffraction examination gives $\angle OOO$ as $127 \pm 3°$, $O, O = 1\cdot26 \pm 0\cdot02$.* These results do not, however, confirm the diagram $O{=}O{=}O$ (I), for, in view of the established tetrahedral nature of the quadrivalent oxygen bonds, this molecule would be linear, like CO_2. The valency angle $125°$ is that between a double bond and a single bond (as in $COCl_2$, p. 69), and the structure indicated is $O{=}O$—O. In this not only is univalent oxygen unacceptable, but the molecule would take the form of a scalene triangle (with sides $1\cdot20$ and $1\cdot46$), whereas the physical evidence indicates an isosceles triangle.† Here again we reach a difficulty only resolvable by appeal to modern electronic theory (p. 134).

(b) *Sulphur*

Like oxygen, sulphur exerts bivalency in numerous compounds, but its higher valencies of 4 and 6 give rise to very stable and important compounds.

(i) *Bivalent sulphur*. Sulphur is clearly bivalent in H_2S, SCl_2, as well as in the elementary molecule S_8 (p. 42), and in S_2Cl_2. The constitution of this latter chloride has been as much debated as that of hydrogen peroxide, and for similar reasons, for its chemical reactions (e.g. hydrolysis) suggest the structure $S{=}S{\Big\langle}_{Cl}^{Cl}$ rather than the symmetrical form Cl—S—S—Cl. The latter has, however, been firmly supported by examination under electron-diffraction.‡ S, S is $2\cdot05 \pm 0\cdot03$, close to that in the molecule S_8 ($2\cdot08$), and $\angle ClSS = 103 \pm 2°$ is also in agreement with that in S_8 ($105°$). In SCl_2 (prepared by the prolonged action of chlorine upon S_2Cl_2) $\angle ClSCl$ is the same ($101 \pm 4°$), and $S, Cl = 1\cdot99 \pm 0\cdot03$. In view of the stability of O_2 and S_2 (in the vapour of sulphur) it is surprising that the oxide SO (prepared from $SOCl_2$ by contact with metals) is so unstable, yielding S and SO_2 even at room

* Shand and Spurr, *J. Amer. Chem. Soc.* **65**, 179 (1943).
† On electric moment of O_3, see Lewis and Smyth, *J. Amer. Chem. Soc.* **61**, 3063 (1939).
‡ K. J. Palmer, *J. Amer. Chem. Soc.* **60**, 2360 (1938)—much of the physical data in the succeeding paragraphs are drawn from this valuable memoir.

temperature.* It is clear that the bivalency of both oxygen and sulphur gives rise to stable compounds only when it is exerted towards *electropositive* elements or groups.

(ii) *Quadrivalent sulphur.* The stereochemistry of the sulphonium salts, and of the sulphoxides and sulphinic esters, has already been touched upon (p. 34). The isolation in optically active forms of the sulphonium ion

$$(CH_3.C_2H_5.S.CH_2COOH)^+$$

by Pope and Peachey in 1900† provided the first demonstration of the pyramidal form of the complex Sabc (pp. 33, 34).

The X-ray analysis of zinc blende (ZnS) shows sulphur in tetrahedral configuration like oxygen in ZnO (see p. 81). Quadrivalency is shown directly in SF_4 (b.p. $-40°$) and in SCl_4 (stable only at low temperatures); there is some evidence that the latter is to be regarded as a type of sulphonium salt $(SCl_3)^+Cl^-$.‡ The structures of sulphur dioxide and thionyl chloride $SOCl_2$ have been elucidated by physical methods by Schomaker and Stevenson,§ Pauling and Brockway,‖ Cross and Brockway¶ and by Palmer.** In both compounds S, O is 1.45 A. $SOCl_2$ resembles the sulphoxides in being pyramidal (S apical), with $\angle OSCl = 106°$ and $\angle ClSCl = 114°$; S, Cl $= 1.99$ as in SCl_2. In $SOBr_2$, $\angle BrSBr = 96 \pm 2°$.†† $SOCl_2$ is thus quite unlike the planar $COCl_2$ (p. 69). In SO_2, $\angle OSO$ is $119.5°$ (cf. CO_2). The $S{=}O$ bond-length may be computed as follows: for C, S in CS_2 Cross and Brockway†† give 1.54 ± 0.03. We assume the relation

$$S{=}O = C{=}S - \frac{C{=}C}{2} + \frac{O{=}O}{2}$$

$$= (1.54 \pm 0.03) - 0.665 + 0.60 = 1.475 \pm 0.03.$$

---

* Cordes and Schenk, *Z. anorg. Chem.* **214**, 33 (1933); Kassel and Montgomery, *J. Chem. Physics*, **2**, 417 (1934).

† *J. Chem. Soc.* **77**, 1072.

‡ Lowry and Jessop, *J. Chem. Soc.* 1931, p. 323; see also $TeCl_4$ below.

§ *J. Amer. Chem. Soc.* **62**, 1270 (1940).

‖ *J. Amer. Chem. Soc.* **57**, 2684 (1935).

¶ *J. Chem. Physics*, **3**, 821 (1935).

** *Loc. cit.* supra.

†† Stevenson and Cooley, *J. Amer. Chem. Soc.* **62**, 2477 (1940).

‡‡ *Loc. cit.* supra.

By adding the radii S— = 1·04 A. (as in S_8) and O— = 0·73 (as in $S_2O_3^=$) we find the bond-length of S—O = 1·77 A. The data therefore indicate the bond-diagrams $O{\nearrow}^{S}{\searrow}O$ and $O{\nearrow}^{S}{\searrow}_{Cl}^{Cl}$.

As thionyl chloride is the chloride of sulphurous acid, we should predict for the sulphite ion $SO_3^=$ the structure $\left[{}_{O}{\nearrow}^{S}{\searrow}_{O}^{O} \right]^=$, pyramidal in shape (S apical). X-ray analysis of Na_2SO_3* confirms the pyramidal form, and gives the height of S above the plane of the O atoms as only 0·51 A., but shows that the three S, O distances are all equal, the distance being 1·39 A., which is notably less than S, O in SO_2 or $SOCl_2$ (p. 84 and *p. 135*).

(iii) *Sexavalent sulphur.* Sexavalency is indubitable in the exceedingly stable gas SF_6 (sublimes at − 63·8°); it is remarkable that SF_4 is rapidly hydrolysed by water and attacked by aqueous alkali, while SF_6 is quite unaffected even by fused alkalis. The volatile form of SO_3 has been examined by the method of electron diffraction.† The S, O distances are all equal, and the same as in SO_2 (1·43 ± 0·02 A.), and ∠OSO is 120°. The molecule is therefore in the form of an equilateral triangle with the central S atom doubly bonded to oxygen at the apices, a configuration confirmed by its zero electric moment.‡ Sulphuryl chloride, SO_2Cl_2, forms an irregular tetrahedron of the structure ${}_{O}{\searrow}^{O}S{\searrow}_{Cl}^{Cl}$. The $SO_4^=$ ion is found by X-ray analysis of sulphates to be a regular tetrahedron with S, O 1·51 (cf. *p. 154*).

The stereochemistry of sulphur

Since in exerting the maximum valency an octahedral disposition of bonds results, it is of interest to examine how the bond dispositions are derived from the octahedral configuration when double bonds arise, and when bonds are suppressed in the lower valencies. In fig. 24 the six geometrically-equivalent positions a to f represent the positions of F in SF_6. Positions labelled 1, 2

* Zachariasen and Buckley, *Phys. Rev.* 37, 1295 (1931).
† Palmer, *loc. cit.*
‡ Smits, Moerman and Pathuis, *Z. physikal. Chem.* B, 35, 60 (1937).

and 3 lie upon the bisectors of the angles between the octahedral bonds, and will be occupied by O in SO_3. The geometry of the octahedron requires that positions 1, 2, 3 and S lie in one plane, as is actually found in SO_3 (see above). In SO_2Cl_2 oxygen atoms will occupy two of the positions 1, 2 and 3, and Cl the remaining two positions (e.g. O at 1 and 2, Cl at e and d). Such an arrangement gives $\angle OSO = 120°$ and $\angle ClSCl = \angle OSCl = 90°$. The experimental values are 119·5°, 111° and 106° respectively.

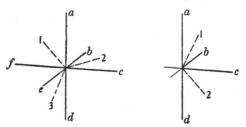

Fig. 24. The stereochemistry of S^{VI} and S^{IV}.

To arrive at the configuration of the quadrivalent sulphur atom we expect to have to suppress two of the positions a to f. Such a reduction in the number of positions may clearly be effected by suppressing either a pair of *polar* positions (e.g. a and d), or a pair of *vicinal* positions (e.g. a and b). The former procedure leads to a square configuration for S^{IV}; SO_2 would be linear, and $SOCl_2$ planar. As these deductions prove incorrect (see above), it may be assumed that the S^{IV} configuration is that shown in fig. 24. The instability of SCl_4 prevents its use to confirm this configuration, but the structure of $TeCl_4$* agrees with it, as do also those of $(C_6H_5)_2SeBr_2$ and $(C_6H_5)_2SeCl_2$.†

In SO_2, oxygen atoms are at positions 1 and 2, with $\angle OSO = 120°$, as found experimentally. In $SOCl_2$, oxygen is at 1 (or 2), and Cl at c and d (or a and b). Theoretically $\angle ClSCl$ and $\angle OSCl$ are both 90°; the actual angles measured are 114° and 106° respectively. In the macromolecular lattices of some metallic sulphides (e.g. ZnS) the regular tetrahedral arrangement (S central) is

* Stevenson and Schomaker, *J. Amer. Chem. Soc.* 62, 1267 (1940).

† Hamburger and McCullough, *J. Amer. Chem. Soc.* 63, 803 (1941); 64, 508 (1942).

adopted (fig. 6, p. 41). The proved optical activity (and hence dissymmetry) of the sulphonium salts (see above) requires that the anion occupy position a or d, but not b or c.

(c) *Selenium and tellurium*

Like sulphur, both these elements show a minimum valency of 2 (as in H_2Se and H_2Te), and a maximum valency of 6 is shown in SeF_6 (sublimes $-46\cdot8°$), TeF_6 (sublimes $-38\cdot9°$) and in telluric acid which is hexabasic and has the true ortho-form $Te(OH)_6$. Quadrivalency is shown in SeF_4 (b.p. $90\cdot5°$) and in $TeCl_4$, as well as in the oxides SeO_2 and TeO_2. Both these solid oxides, however, show chemical and physical properties illustrating the expected gradation in the group. SeO_2 acts as a moderately powerful *oxidizing* agent, increasingly used in organic chemistry (see account in *Ann. Reports*, 1937, p. 238). Its crystal structure consists of long macromolecular chains as shown:

$$
\begin{matrix}
& O && O \\
& \| && \| \\
-Se & -O-Se-O- & Se- \\
& \| \\
& O
\end{matrix}
$$

$\angle\,SeOSe = 125°,$
$\angle\,{-}OSeO{-} = 98\pm2°,$
$\angle\,{-}OSeO{=} = 90°.$*

TeO_2 has the structure of brookite, one of the three polymorphic forms of TiO_2, the others being rutile and anatase. All the forms are considered to be ionic lattices, and thus quadrivalent Te has claims to be regarded as at least partially metallic, a character further supported by the instability of H_2Te, and by the physical characters of the tetrahalides $TeCl_4$, m.p. 224°, b.p. 414°; $TeBr_4$, m.p. 390°, b.p. 420°; fused $TeCl_4$ has the high electrical conductivity of a salt (cf. SCl_4 above).

GROUP VII (B). THE HALOGENS

From the relation in stability of Cl_2O_7 and I_2O_5, the instability of the fluoride IF_7, and the possible existence of salts of cationic iodine, we see the regular tendency in B families towards lowered principal valency and increased metallic properties. Unusual features include the persistence of the univalency of fluorine, and the absence of stable oxides of bromine.

* McCullough, *J. Amer. Chem. Soc.* **59**, 789 (1937); *Ann. Reports*, 1937, p. 160.

(a) *Fluorine*

All interhalogen compounds containing only one F atom are binary (e.g. FCl, FBr). The only oxide known to be stable at room temperature is F_2O (b.p. $-146°$), although another, F_2O_2, has been reported as existing at low temperatures. F, O in F_2O is $1·38 \pm 0·03$, and $\angle FOF = 101 \pm 1·5°$*. Some doubt exists on the exact value of the radius of F in combination, but taking a mean value of $0·64$ and $O— = 0·73$ (as in $S_2O_8^=$) the predicted length of F—O is $1·37$.

It has long been recognized that hydrofluoric acid, unlike the other halogen hydracids, functions in concentrated aqueous solution as a weak, monobasic acid $H(HF_2)$, and is strongly associated in the anhydrous vapour state. In crystals of the hydrofluorides MHF_2 the group F_2 occurs, and the F, F distance obtained by X-ray analysis appears to be variable:

NaHF₂ $2·5 \pm 0·2$ Pauling, *Z. Krist.* **85**, 380 (1933).

NH₄HF₂ $2·32 \pm 0·03$ $\left\{ \begin{array}{l}\text{Helmholz and Rogers, } J.\ Amer. \\ Chem.\ Soc.\ \textbf{61},\ 2590\ (1939)\ \text{and }\textbf{62}, \\ 1533\ (1940).\end{array} \right.$

KHF₂ $2·26 \pm 0·01$

Recent studies by the method of electron-diffraction† confirm earlier estimates of the far-reaching extent of the polymerization in the vapour, and suggest polymers ranging in composition from $(HF)_2$ to $(HF)_5$. The polymers appear to be of zig-zag form, with F, F $= 2·55 \pm 0·03$, and $\angle FFF = 140 \pm 5°$.‡ The crystalline

hydrogen halides HCl, HBr and HI all have cubical or nearly cubical unit cells, face-centred, and containing 4 molecules per cell. Solid hydrogen fluoride (at 91° K.) has a *tetragonal* unit cell, with the very disparate edges $a = 5·45$, $c = 9·95$, and containing 16 molecules.§ The dimensions are compatible with a structure of long zig-zag chains, —H—F—H—F—, in which

* Bernstein and Powling, *J. Chem. Physics*, **18**, 685 (1950).

† Bauer, Beach and Simons, *J. Amer. Chem. Soc.* **61**, 19 (1939).

‡ See also Long, Hildebrand and Morrell, *J. Amer. Chem. Soc.* **65**, 182 (1943).

§ Günther, Holm and Strunz, *Z. physikal. Chem.* B, **43**, 229 (1939).

$F, F = 2\cdot7$ and $\angle F, F, F = 134°$, close to the values reported for the gaseous state. (H cannot be located by X-ray diffraction (p. 117).)

There is clearly something very abnormal about the F, F linkage in both the ion HF_2^- and in the polymerized acid, for the distance F, F is about double the known F—F distance, and does not appear to allow a direct F—F bond. The fact, proved by X-ray analysis of KF, that treatment of KHF_2 with KOH furnishes the normal halide KF and not K_2F_2 may be regarded as further evidence against the existence of a direct F—F bond in the ion HF_2^-. The difficulty can hardly be said to be solved by proposing the structure $(F—H—F)^-$ for the ion, for we seem to evade bivalent fluorine only to accept bivalent hydrogen. The application of modern principles has much elucidated this dilemma (*p. 222*). The chemical facts may be summarized by saying that no clear evidence exists for assigning a valency greater than 1 to fluorine.

(b) Chlorine and bromine

The properties of the chief oxides of chlorine, which play an important part in assessing the valencies of chlorine, are summarized below:

	Cl_2O	ClO_2	Cl_2O_6	Cl_2O_7
M.p. ° C.	− 116	− 59	3·5	− 91·5
B.p. ° C.	2	11	(200)	80
V.p. at 0° (mm.)	699	490	0·31	23·7
Heat of formation (Cal.)	18·04	23·5	?	63·4

The endothermicity of the oxides, reckoned per oxygen atom, decreases with increasing oxygen, probably to a minimum at the heptoxide, which is thus the least unstable of the series.

For the gaseous Cl_2O we have

$$Cl, O = 1\cdot70 \pm 0\cdot02, \quad \angle ClOCl = 111 \pm 1°.*$$

The bond-length of Cl—O from Table 6 is $0\cdot99 + 0\cdot73 = 1\cdot72$. The bond-diagram Cl—O—Cl is thus confirmed, with univalent chlorine as expected. The fact that the hypohalous acids can act

* Dunitz and Hedberg, *J. Amer. Chem. Soc.* **72**, 3108 (1950).

both as agents of oxygenation and of halogenation, expressible respectively in the equations

(i) $HClO + R = HCl + RO$,   (ii) $RH + HOCl = H_2O + RCl$,

led to the conception that these acids existed in tautomeric equilibrium, e.g. $HClO \rightleftharpoons HOCl$. A deeper understanding of the mechanism of chemical reaction has not however lent support to such views, which, owing to the instability of the acids, would in any event be incapable of direct proof.

Acids HX as a rule add to the ethylenic bond as the addenda H and X. Hypochlorous acid is exceptional in giving chlor-hydrins, by the addenda OH and Cl, but in this it is analogous to NOCl, which adds as NO— and Cl—. The first-formed organic product of the action of chlorine on ethyl alcohol is the hypochlorous ester C_2H_5OCl, which then rapidly reacts with excess alcohol to give acetaldehyde and HCl, but not C_2H_5Cl, as primary products.[*] The reaction no doubt proceeds in the stages

$$C_2H_5OCl + HO . C_2H_5 \rightarrow C_2H_5\!-\!O\!-\!O\!-\!C_2H_5 + HCl,$$
$$C_2H_5\!-\!O\!-\!O\!-\!C_2H_5 \rightarrow CH_3CHO + C_2H_5OH.$$

It is difficult to understand how these products, particularly HCl, could be smoothly obtained if the ester had the constitution C_2H_5ClO.

As in the case of the lowest valencies of O and S, those of the halogens Cl and Br are exerted far more stably towards electro-positive elements and groups than to electronegative.

The oxide Cl_2O_6, prepared by oxidizing ClO_2 with ozone, and whose molecular weight is established cryoscopically in CCl_4 and in cyclohexane, dissociates completely to ClO_3 on vaporization.[†] It is informative to compare the physical properties of the oxides and ions ClO_2, ClO_3^- and ClO_3 with those of the neighbouring element sulphur which are analogous in formula:

	M.p.	B.p.	Aq. sol. at 18°	Mol. vol.	Bond-angle	X, O	X—O
SO_2	−72·5°	−10°	42 vol.	44·4	119·5°	1·45	1·63
ClO_2	−59°	11°	47 vol.	40·5	ca. 120°‡	1·49‡	1·70

‡ Dunitz and Hedberg, *J. Amer. Chem. Soc.* **72**, 89 (1950).

* Chattaway and Backeberg, *J. Chem. Soc.* **125**, 1097 (1924).
† Goodeve and Richardson, *J. Chem. Soc.* 1937, p. 294.

Neither gas shows association in the gaseous state or upon condensation.

		$\angle$OXO	X, O	
SO_2	Polymerizes on condensation	M.p. 16·8°		Planar shape
ClO_2	The only oxide of Cl to polymerize on condensation	M.p. 3·5°		?
SO_3^-	Pyramidal, S apical	107·5°	1·39*	
ClO_3^-	Pyramidal, Cl apical	109°	1·48†	

* Zachariasen and Buckley, *Phys. Rev.* **37**, 1295 (1931); see also p. 85.
† Zachariasen, *Z. Krist.* **71**, 501 (1929).

The data appear sufficiently analogous to suggest similar formulae for the two sets of oxides. The chlorine oxides, however, in their reactions with water or alkalis differ markedly from those of sulphur: upon hydrolysis the dioxide yields the ions ClO_2^- and ClO_3^-, while the trioxide gives a mixture of ClO_3^- and ClO_4^-. Chlorine exhibits tervalency only in the fluoride ClF_3, a colourless substance, b.p. 12·1°.

Cl_2O_7. This oxide, which is the least unstable oxide of chlorine, (p. 89) is prepared by the dehydration of perchloric acid and again gives rise to the acid in contact with water. In the perchlorates the ClO_4^- ion is known from X-ray analysis to be tetrahedral, with central Cl atom; the Cl, O distance is 1·46, slightly less than in ClO_3^- (1·48) and in ClO_2 (1·49). These data suggest for the anhydride the following structure:

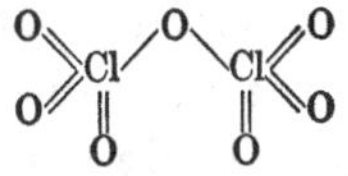

with septivalent chlorine. The electric moment of the oxide is found to be 0·72 D.‡ The moment of Cl—O is estimated to be 0·73 D.§ by Pauling;∥—that of Cl=O is unknown, but may be taken to have a value not greater than that of S=O, which is 1·6 D., the total moment of SO_2. Assuming that O surrounds the Cl tetrahedrally in the anhydride, and the tetrahedral angle is maintained for $\angle$ClOCl, we calculate a resultant moment of about 0·8 D.

‡ Fonteyne, *Amer. Abstr.* **33**, 1214 (1939).
§ D. represents Debye units, see p. 37.
∥ *Nature of the Chemical Bond*, p. 68.

From the above summary of important Cl compounds it appears that Cl exhibits valencies between 1 and 7 inclusive except 2, but the extreme values of 1 and 7 are found in its most stable compounds.

Bromine appears to form no oxides stable at room temperature, although Br_2O may exist at low temperatures. The acids HBrO and $HBrO_3$ are well known, but there appears to be no perbromic acid $HBrO_4$. The fluoride BrF, b.p. ca. 20°, is unstable and rapidly undergoes disproportionation to Br_2 and BrF_3 (b.p. 127°). The highest fluoride is BrF_5 (b.p. 40·5°). As is not unusual the highest fluoride is the most volatile. The valencies of bromine may be taken to be 1, in HBr, HOBr, and (unstably) in BrF and Br_2O; 3 in BrF_3 alone (cf. ClF_3); 5 in BrO_3^- and BrF_5.

(c) *Iodine*

Much of the varied chemistry of iodine has been aptly interpreted and systematized by assuming the element can exist in electropositive states, such as I^+ and I^{+++}. It should not, however, be overlooked that while an acceptable reaction mechanism may require a reacting molecule to be transiently in an ionized state this conclusion is no safe guide to the constitution of its *permanent* (or 'resting') molecular state. Experimental facts may appear to require that bromine at the moment of its addition to ethylene be in an ionized state $Br^+.Br^-$, but there is equally no doubt that in the permanent state of the bromine molecule the atoms are completely equivalent, as expressed in the formula Br—Br. In general only a direct examination, usually by physical means, of a compound, necessarily in its resting state, can lead to a proof of its constitution. A summary of the valencies exerted is found in the halides: ICl and IBr; ICl_3; IF_5 and IF_7. A fluoride lower than IF_5 has not been isolated.

(i) *Univalent iodine*. The hydride HI is distinguished from the other halogen hydrides by its positive heat of formation (6·40 Cal./mol. at room temperature), and its consequent instability, represented by the easy dissociation at moderate temperatures, and the high reducing potential. The reduction of iodate in the presence of excess of hydrochloric acid (as in the well-known

titration method of Andrews) can be explained by the successive
equations:

$$IO_3^- + 4H^+ + 4e = IO^- + 2H_2O, \qquad (1)$$

$$IO^- + H^+ = IOH, \qquad (2)$$

$$IOH + HCl = ICl + H_2O, \qquad (3)$$

$$ICl + Cl^- = ICl_2^-,$$

of which (3) suggests an amphoteric character for hypoiodous
acid, with the corollary that ICl should show some salt-like
properties. For this requirement, however, the recent examination of ICl by X-ray diffraction affords no support (p. 95). In
Lang's method of titration with iodate an excess of the very
weak hydrocyanic acid successfully replaces hydrochloric acid,
leading to the quantitative formation of ICN as final product.
This is not easy to relate to IOH as a hydroxide when it is recalled that both the acid and base functions of amphoteric compounds must necessarily be weak. It has, however, recently been
confirmed that *liquid* ICl* and its solutions in acetic acid and
nitrobenzene conduct electrolytically, with cathodic deposition
of iodine.†

(ii) *Tervalent iodine.* The trichloride ICl_3, prepared by the
direct union of Cl_2 with I_2 or ICl, has the yellow colour of all
tervalent iodine compounds. Like ICl, the trichloride is electrically conducting in the liquid state,‡ but its recently explored
structure, although unexpected and difficult to explain, certainly offers no indication of simple I^{+++} ions (see p. 95). The
complex salt $KICl_2$ is obtained by passing chlorine in limited
amount through a suspension of iodine in a solution of KCl. By
oxidizing iodine in the presence of suitable reagents (e.g.
phosphoric or perchloric acids, acetic anhydride) numerous
compounds have been prepared, including IPO_4, $I(ClO_4)_3$,
$I(CH_3COO)_3$, and $IO.NO_3$, which from their formulae might be
considered as salts of the tripositive cation I^{+++}, or of the oxy-
cation IO^+. The yellow crystalline oxides of composition I_2O_4

* Greenwood and Eméleus, *J. Chem. Soc.* 1950, p. 987.
† Sandonnini and Borgello, *Atti Accad. Lincei*, **25**, 46 (1937).
‡ Greenwood and Eméleus, *loc. cit.*

and I_4O_9 have been widely regarded as $IO^+ . IO_3^-$ and I^{+++}. $(IO_3)_3^-$ respectively.* Until the crystal structures of such compounds have been determined such interpretations cannot be regarded as finally established.

(iii) *Quinquevalent iodine.* The pentafluoride IF_5 (b.p. 97°), which is rapidly attacked by water, probably with the initial production of IOF_3,† shows clear quinquevalency. The colourless, solid, and wholly acidic oxide I_2O_5 is prepared by the direct oxidation of elementary iodine by a variety of agents: aqueous chlorine or bromine, hypochlorites, or $KMnO_4$; it may also be obtained by the dehydration of HIO_3 (produced by oxidizing I_2 with HNO_3), and gives rise again to the acid on treatment with water. Technical difficulties, associated with the great difference in scattering power between I and light atoms such as O and F, have unfortunately prevented a successful X-ray or electron-diffraction examination of I_2O_5 and IF_5 (see below), but there is no reason to doubt quinquevalency in the pentoxide.

The contrasted reactions (a) $2HI + Cl_2 \rightarrow I_2 + 2HCl$ and (b) $2HClO_3 + I_2 \rightarrow 2HIO_3 + Cl_2$ illustrate the great difference between I and Cl in relative affinity for oxygen and hydrogen. The structure of iodic acid and the iodates is considered in the special concluding section.

(iv) *Septivalent iodine.* The gaseous fluoride IF_7 (m.p. 4·5°; b.p. 5–6°), although readily dissociated thermally to IF_5 and F_2, is only slowly attacked by water at room temperature (cf. SF_6, SF_4 and IF_5 above). The corresponding ortho-acid $I(OH)_7$ gives rise to many anhydro-acids, such as H_5IO_6 (para-periodic acid) and HIO_4 (meta-periodic acid), as well as more complex acids derived by dehydration from more than one molecule of the ortho-acid, which is itself unstable. The stable existence of the salt Ag_5IO_6, and the results of crystallographic analysis given below, confirm the formula for para-periodic acid, which however is not more than tribasic in most of its reactions. Unlike $HClO_4$ in its relation to $HClO_3$, per-iodic acid is a powerful oxidizing

* Masson, *Nature*, **139**, 150 (1937); Partington and Bahl, *J. Chem. Soc.* 1935, p. 1258.

† Ruff and Braida, *Angew. Chem.* **47**, 480 (1934).

agent, much stronger than HIO_3. The term 'per-iodic' is however a misnomer, in that the acid does not contain the true peroxide grouping —O—O—.

Summary of investigations of the structures of compounds of iodine

ICl and ICN. In being a packing (in chain formation) of ICl molecules (I, Cl = 2·40), the structure of the monochloride does not differ essentially from that of molecular iodine, but the *intermolecular* distances I, I and I, Cl are shorter than expected: I, I in I_2 is 3·54, and in ICl it is 3·08.* ICN is composed of linear molecules I—C≡N.

ICl_3. The solid is composed of symmetrical, planar and dimeric molecules I_2Cl_6 of the form

$$\begin{array}{ccc} Cl & Cl & Cl \\ & \diagup & \diagup \\ & \!\!\!\diagdown I \diagdown & \!\!\!\diagdown I \diagdown \\ Cl & Cl & Cl \end{array}$$

The terminal I, Cl distances are 2·38 as in ICl, but the four central bonds are longer at 2·70. All bond angles are close to 90°. It is plausible that in *liquid* ICl_3 such molecules could to some extent give rise to the ions ICl_4^- and ICl_2^+, the latter being the analogue of the oxy-cation IO^+, but the properties of the *crystal* preclude the presence of discrete ions.†

IF_5. Infra-red absorption spectra and the related Raman spectra (see section 4, p. 39) arise from intramolecular vibrations, the number and type of which depend upon molecular symmetry: molecules of similar shape have analogous spectra. The spectra of IF_5 show that the molecule possesses a fourfold axis of symmetry, a result that alone suffices to prove that one I—F bond must lie on that axis and the other four be directed radially from it to the corners of a square. The position of the iodine atom is probably somewhat above the plane of the radially placed fluorine atoms.‡

$(NH_4)I_3$, $CsICl_2$ and $KICl_4$. These salts contain the anions I_3^-,

* Boswijk, van der Heide, Vos and Wiebenga, *Acta Cryst.* **9**, 174 (1956).
† Boswijk and Wiebenga, *Acta Cryst.* **7**, 417 (1954).
‡ Lord *et al.*, *J. Amer. Chem. Soc.* **72**, 522 (1950).

ICl_2^- and ICl_4^-, of which the triatomic are linear* with central I in ICl_2^-, while in the last all atoms are co-planar, and the I atom is at the centre of a square of Cl atoms.† I, Cl = 2·34 in both chloro-ions.

KIO_2F_2. In this interesting salt, prepared by treating KIO_3 with concentrated hydrofluoric acid, the anion $IO_2F_2^-$ is found to be a discrete unit of the structure, and its configuration, shown in fig. 25, is reminiscent of that of S^{IV}, and Te^{IV} in $TeCl_4$ (p. 86). The fluorine and iodine atoms are linearly arranged: $\angle OIO$ = ca. 100°, and the plane of OIO is perpendicular to the

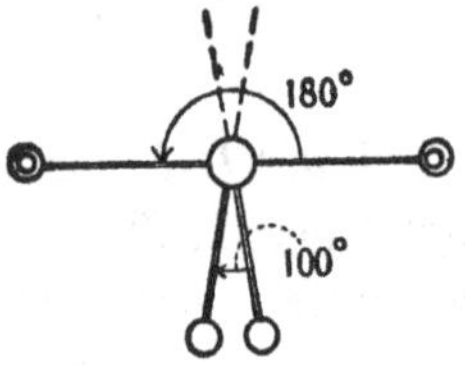

Fig. 25. The stereochemistry of $(IO_2F_2)^-$.
○, I; ◎, F; ○, O.

FIF axis. I, O = 1·93 ± 0·05, and I, F = 2·00 ± 0·05.‡

From ICl_4^- above we have I—Cl = 2·34; taking Cl— = 0·99, F— = 0·64 and O— = 0·73, we calculate for I—O 2·08 and for I—F 1·99.

As I, O distances of ca. 1·8 are found in HIO_3, which are associated with I, O double bonds, we may assume that in $IO_2F_2^-$ all four atoms are singly bonded to the central iodine. By giving one negative charge to each O atom and one positive charge to the I atom we place O in the OH^- condition, and restore normal bivalency to oxygen, while maintaining the single negative charge upon the whole ion.

$(NH_4)_2H_3IO_6$ (di-ammonium tri-hydrogen para-periodate). The point of interest arising from X-ray analysis§ is that each I atom is surrounded by a (nearly) regular octahedron of O atoms, each of whose distances from the central iodine is 1·93 ± 0·03, as found for I, O in $IO_2F_2^-$, and corresponding with I—O. Bivalency is restored to oxygen by giving each atom one negative charge: to maintain the net charge of the IO_6^{5-} ion one positive charge must again be placed on the central iodine.

The iodates MIO_3. The difficulty experienced in diffraction methods of accurately placing light elements in combination

* Mooney, Z. Krist. 90, 143 (1935).
† Mooney, ibid. 98, 377 (1938).
‡ Helmholz and Rogers, J. Amer. Chem. Soc. 62, 1537 (1940).
§ Helmholz, J. Amer. Chem. Soc. 59, 2036 (1937).

with iodine (see above) no doubt explains the earlier misconceptions of the constitution of crystalline iodates, which were supposed to have the perovskite structure. This cubical arrangement (fig. 26, with broken line spheres) is found, not only in

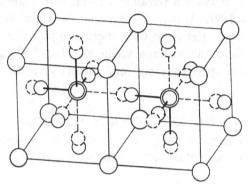

Fig. 26. The structure of iodates, MIO_3 (idealized), showing the relation to two unit cells of the perovskite structure. ◯, M; ○, O in MIO_3; ◌, O in perovskite; ◎, I.

perovskite itself, $CaTiO_3$, but in numerous compounds of analogous general composition ABO_3, and because of its high symmetry, and the measured distances between the crystal units, has been acceptably interpreted as an ionic structure built of the cations A^{n+} and B^{m+} with oxygen ions $O^=$: the cations A and B may be very various, always provided that $n+m = 6$. The units in crystalline iodates MIO_3 were therefore assumed to be M^+, I^{+++++} and $O^=$, in extreme contrast with chlorates and bromates which were known to contain discrete anions ClO_3^- and BrO_3^- of pyramidal shape. A re-investigation of $NaIO_3$* has proved that while the positions of M^+ and I are very nearly those in the perovskite structure, the true disposition of the oxygen atoms are obtained by slight but significant displacements from the face centres in the cubical structure (whole spheres in fig. 26). In the structure thus produced (which has only orthorhombic symmetry) discrete IO_3^- ions, with I, O 1·81, and $\angle OIO = 94°$ (cf. HIO_3 below), are at once discerned. Only one minor difference between the halates MXO_3 remains: in chlorates and bromates the cations surround the anions in the NaCl arrangement, but in iodates in the CsCl form (figs. 9 and 11).

* MacGillavry and van Eck, *Rec. trav. Chim.* 62, 729 (1943).

HIO_3. The orthorhombic α-form of the acid has been examined in detail.* The results indicate that the crystals are built essentially of HIO_3 molecules, although the method of X-ray analysis cannot directly discover the position of H atoms. The IO_3 group is prominent; it takes a pyramidal form, with I apical, and I, $O_I = I$, $O_{II} = 1·80$; I, $O_{III} = 1·89$: the H atom is probably attached to O_{III}. The three OIO angles are probably not quite equal, but have the average value $98 \pm 3°$ (cf. ClO_3^-, p. 91). These data are approximately conformable with the bond-diagram $HO—I{<}{\atop{}}^O_O$.

The meta-periodates MIO_4. The crystal of $NaIO_4$ contains well-marked tetrahedral ions IO_4^- similar to ClO_4^- in the perchlorates.† I, O = 1·79, a value close to the two equal I, O lengths in HIO_3 (see above).

The ascertained configuration of the ions ICl_4^- and $IO_2F_2^-$ throws some light on the probable configuration in a non-ionized compound of I^V such as IF_5. An ion IX_4^- may be regarded as stereochemically derived from IX_5 by rendering one of the five bonds in IX_5 latent. If we attribute to IX_5 the shape of a trigonal bi-pyramid (fig. 13), then the equatorial angles XIX must be 120°, and if one of these bonds is latent the remaining angle XIX will be at least 120°. Actually, however, the (equatorial) angle OIO in $IO_2F_2^-$ is 100°, and therefore cannot be derived in the way suggested from a trigonal bi-pyramid. An angle of about 100° is, however, that which might reasonably be expected if IX_4^- were derived from an *octahedral* configuration IX_6 in which *two* equatorial bonds were latent (fig. 25). The square configuration of ICl_4^- is explained if we assume that in this case both *polar* bonds of the octahedral parent form are latent. In this connection it is significant that the octahedral arrangement of IX_6 is actually found in the para-periodates. We may therefore conclude that the configuration of IX_5 is derivable from a regular octahedron, with one bond latent (cf. p. 95).

---

* Helmholz and Rogers, *J. Amer. Chem. Soc.* **63**, 278 (1941).
† Hazlewood, *Z. Krist.* **98**, 439 (1938).

Vanadium, chromium and manganese

Although these three metals ostensibly belong to Groups V, VI and VII respectively, and show in their highest oxides and halides the appropriate valencies of 5, 6 and 7, in being definitely metallic and in forming compounds of types very different from those of the other group-elements, they are sharply distinguished from the typical members of their groups. For this reason and also because they provide an excellent example of conditions in the middle of a long series, they will be here considered together. In the following tables, which very briefly summarize the valency relationships, an attempt has been made to indicate by arrows changes that readily occur: arrows on the same level indicate simultaneous changes.

Table 15. *Vanadium* (Group V A)

Valency	2	3	4	5
Exerted as	$V^{++} \longrightarrow V^{+++}$		(V^{++++}) $\downarrow$ VO^{++}	$(VO_4^{\equiv})$ $\downarrow$ $(VO_3)_4^{\equiv}$
Colour	Violet	Green	Blue	Colourless
Examples		$VF_2 \longleftarrow$ heated $(200°)$ $VF_4 \longrightarrow$ heated VF_5 (Colourless solid, b.p. 111°) $VCl_2 \longrightarrow VCl_3 \longleftarrow VCl_4$ (m.p. $-28°$, b.p. 150°) $\longrightarrow VO.Cl_2 \longleftarrow$ aq. $VO.Cl_3$ (m.p. $-77°$, b.p. 127°) aq. $VO \longrightarrow V_2O_3$ aq. $VO.O$ V_2O_5 (red)		

In the mere numerical values (3 and 5) of its most stable valencies (in anhydrous conditions), vanadium shows relationship with the B members of Group V. In all its 5-valent compounds the metal is in a non-ionic condition, and in this shows real analogy with the B members. The vanadates and V_2O_5 are weak oxidizing agents resembling the arsenates and As_4O_{10}. Orthovanadates $(VO_4^{\equiv})$, stable only in alkaline solution, easily pass over into metavanadates, with condensed anion $V_4O_{12}^{\equiv}$ resembling tetrametaphosphate $P_4O_{12}^{\equiv}$, but V_2O_5 has a structure very different from that of P_4O_{10} or As_4O_{10}. Each V atom is surrounded by an (irregular) tetrahedron of O atoms, and the tetrahedra are linked together at three corners by three of the O atoms (fig. 27). By a slight modification of this structure 'fibre' macromolecules, dependent on —V—O—V—O— chains, are readily produced, and have long been recognized as constituents of colloidal solutions of V_2O_5.* Tervalent vanadium compounds are based on the *cation* V^{+++}, whereas 3-valent N and P are based on the instability to water of tervalent *anions* $N^{\equiv}$ and $P^{\equiv}$ ($P^{\equiv} + 3H^+ \rightarrow PH_3$). The oxide V_2O_3 has the ionic lattice of corundum, Al_2O_3.† The 4-valent ion V^{4+} and the non-ionic halides of 4-valent vanadium are unstable to water and heat respectively,

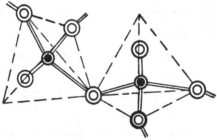

Fig. 27. The structure of V_2O_5. ⊙, V; ◉, O.

as is well shown by the immediate and complete hydrolysis of VCl_4 to the stable vanadyl chloride $VO.Cl_2$, which still contains 4-valent V in the cation VO^{++}, and by the disproportionation of VF_4 on heating (see below). The chloride VCl_4 dissociates on heating to VCl_3 and Cl_2. In aqueous solutions, containing necessarily *hydrated ions*, relationships are different. The ions V^{++} and V^{+++} are strong reducing agents, the former readily liberating hydrogen, and both easily pass to V^{4+}, and thence by hydrolysis to the stable VO^{++}, which represents the most stable valency condition of V in *aqueous solution*. We may contrast the reaction in aqueous solution

$$V^{+++} \text{ aq.} + V^V O_4^{\equiv} \text{ aq.} \rightarrow 2V^{IV}O.O \text{ aq.}$$

(green)   (colourless)   (blue)

with the anhydrous reaction $2V^{IV}F_4 \rightarrow V^{III}F_3 + V^V F_5$. The oxide VO has the MgO structure, and therefore consists of V^{++} and

* Ketelaar, *Amer. Abstr.* **30**, 3295 (1936); *Nature*, **137**, 316 (1936).
† Zachariasen, *Amer. Abstr.* **23**, 1790 (1929).

$O^=$. The crystal forms and general chemistry of the compounds of V^{++} resemble those of Fe^{++} (e.g. VSO_4 and $FeSO_4$ are isomorphic).

Table 16. *Chromium* (Group VIA)

Valency	2	3	6
Exerted as	Cr^{++} ——→ Cr^{+++}		$CrO_4^=$ $(Cr_2O_7)^=$
Colour	Blue	(See below)	Yellow, orange
Examples	$CrCl_2$ ——→ $CrCl_3$ Alums CrO ——→ Cr_2O_3 ←	heat	CrO_3 (red) CrO_2Cl_2 (red liquid, b.p. 117°) —— CrO_3 $(Cr_2O_3 . CrO_3 = CrO_2)$

As in the case of vanadium the principal resemblance of chromium to the B members of the group, e.g. sulphur, is in the compounds showing the highest valency of 6. In solutions of chromic salts various complex cations are usually present, in amounts depending on the concentration and the temperature. By determining the proportion of Cl^- ion by means of precipitation with $AgNO_3$ the following have been identified in chloride solutions:[*]

$$(Cr.6H_2O)^{+++} Cl_3 \quad \text{violet,}$$
$$(Cr.Cl.5H_2O)^{++} Cl_2 \quad \text{pale green,}$$
$$(Cr.Cl_2.4H_2O)^+ Cl \quad \text{deep green.}$$

The ion $(Cr.6H_2O)^{+++}$ is presumably also present in chrome alum, which is violet in colour, but yields green solutions on warming with water. Anhydrous chromic chloride (prepared by passing chlorine over a mixture of Cr_2O_3 and carbon) is also violet, but its volatility suggests a non-ionic nature. The green colour of the (ionic) crystals of Cr_2O_3 (corundum structure) is presumably due to the anhydrous ion Cr^{+++}.

As will be understood from Table 17 manganese exhibits only two principal valencies (2 and 7) in compounds other than oxides (see below). As with V and Cr the highest valency is associated with a completely non-ionic state of the metal. The oxide Mn_2O_7 is a non-conductor. It is doubtful whether, with the exception of the dioxide MnO_2, any simple compound

* See further, p. 170.

8

of 4-valent Mn is known, although very unstable deep-coloured solutions are formed when cold concentrated HCl is mixed with MnO_2. Alums containing Mn^{+++} have been prepared, and are

Table 17. *Manganese* (Group VII A)

Valency	2	3	4	6	7
Exerted as	$Mn^{++} \leftarrow$	$Mn^{+++} \leftarrow$	(Mn^{++++})	$MnO_4^{=} \longrightarrow$	MnO_4^{-}
Colour	Pink	Red	?	Green	Purple
Examples	$MnCl_2 \leftarrow$	$MnCl_3 \leftarrow$ Alums (red)	$(MnCl_4)$	$K_2MnO_4 \xrightarrow{\text{acid}}$	$KMnO_4$
	$MnO \leftarrow$	$Mn_2O_3 \leftarrow$	$\xrightarrow{600^\circ} MnO_2$		Mn_2O_7 (liquid at room temp.)

deep red in colour. The existence of a trioxide MnO_3, corresponding with the manganates, is now considered to be improbable. Compounds of bivalent Mn closely resemble the compounds of ferrous iron.

In regard to the relative stability of its valencies the oxides of manganese contradict the evidence of the salts. All oxides when heated in air are finally converted into Mn_3O_4, which resembles Fe_3O_4 in being constituted of the ions Mn^{++}, Mn^{+++} and $O^{=}$ in the necessary ratio $1:2:4$. The colourless hydroxide $Mn(OH)_2$, like $Fe(OH)_2$, is very readily oxidized by air, at normal temperatures. Mn_2O_3, which, unlike the rhombohedral Fe_2O_3 and Cr_2O_3, has cubical crystals, is like them built of Mn^{+++} and $O^{=}$ ions. MnO is also ionic, and like FeO has the MgO structure. MnO_2 (and also TiO_2, SnO_2 and PbO_2) has the rutile (TiO_2) structure. Each Mn is octahedrally surrounded by six oxygen atoms, and each O by three manganese atoms.* Neither the oxygen octahedron nor the triangle of Mn atoms is quite regular, and it might therefore be conjectured that non-ionic links play a part in the crystal, but it has been shown† that the physical properties of the crystal may be successfully calculated by assuming that it is built of Mn^{4+} and $O^{=}$ ions.

* See fig. 21, p. 71.

† Lennard-Jones and Dent, *Phil. Mag.* (7), 3, 1204 (1927)

Chapter IV

AN ELEMENTARY ACCOUNT OF THE SPIN THEORY OF VALENCY AS APPLIED TO LIGHT ATOMS

SECTION I

As we now hold it the electrical theory of matter is a product of recent physical advances, but, looking back from the present day, we can discern many incentives to the establishment of such a theory, notably the proof given in 1874 by Maxwell that light, emitted and absorbed by matter, is an electromagnetic phenomenon. All physicists were forced to adopt new standpoints in regard to both matter and radiation by the extraordinary progress achieved at the close of the last century in a space of less than one decade. The following chronology may be noted:

1895	The first production of Röntgen radiation (p. 116).
1895–1897	The discovery of the electron (Thomson).
1896–1900	The discovery of radioactivity, and of radium (Becquerel, Curie; developed in succeeding years by Rutherford and Soddy).

Following hard on these great experimental overtures came, in the early years of this century, complementary advances in conceptions and generalizations:

1900	Foundations of the quantum theory (Planck) (p. 105).
1911	The conception of the nuclear atom (Rutherford) (p. 104).
1913–1914	Bohr's interpretation of atomic spectra (p. 106); Moseley's law of atomic number (p. 146).

Space will not permit more than the mention of these great antecedents of the modern theories. We must assume for our purposes the accepted fact that all atoms, normally electrically neutral, contain exactly balanced charges of electricity; and that whereas the positive charge is entirely concentrated in one central particle or *nucleus*, the negative charge is subdivided into units, *electrons*, which are common to all atoms. One atom is distinguished from another primarily by the constitution of the nucleus. The *atomic number*, derivable chemically from the order of the periodic system, or independently from Moseley's law, records directly the total number of extra-nuclear electrons, and the equal net positive charge carried by the nucleus. Nuclear constitution may change without alteration of the net positive charge or of the number of extra-nuclear electrons required for balance. By such nuclear changes distinct atoms (*isotopes*) arise, but having an identical number and arrangement of extra-nuclear electrons these usually lack all chemical distinguishing features, which result only from a difference in the number and arrangement of the electrons. Electrons may be extracted from atoms in a variety of ways, ranging in violence from the ready removal by ionization of the outermost units, to the more deep-seated disturbance produced by penetrating X-radiation, and finally to the enormous concentration of energy required in spontaneous or artificial nuclear disintegration, which last usually results in the emergence of other types of particles besides electrons. In view of the known upper limits to the localization of energy in chemical changes it must be inferred that such changes are connected with disturbances and rearrangements of only the outermost or peripheral electrons.

It is therefore pertinent to examine whether there is any apparent correlation between the energy of atomic ionization, i.e. the energy required to release one peripheral electron, and general chemical reactivity and valency. It may be observed that as most of the work of ionization is accomplished in the neighbourhood of the atom (owing to the operation of the inverse square law) the ionization energy (potential) gives a real measure of the atomic stability of the outer electrons. The data in Table 18 appear to answer the enquiry negatively. Indeed, it was hardly to be expected that the *integral* nature of chemical valency would be wholly explained by mere energetics. The explanation must principally lie in the *arrangement* of the electrons (or, more precisely, their orbits) within the atom.

Table 18. *Ionization potentials.*

	H	He	Li	Be	B	C	N	O	F	Ne
Z	1	2	3	4	5	6	7	8	9	10
I_{volts}	13·5	24·4	5·4	9·3	8·2	11·2	14·5	13·6	17·3	21·5
(X), HX		0	1	2	?	4	3	2	1	0
(X), XF	1	0	1	2	3	4	3	2		0

One electron-volt=approx. 23 Cal./g. atom. Z=atomic number.

It is reasonable to assume for atoms to the left of carbon which yield cations stable to water, that the units of positive charge, known to equal the valency, also equal the number of (active) electrons available for chemical reaction. We notice then that two electrons remain inert in He, Li and Be, and the single electron in hydrogen is active. It is further noticeable that the probable valency to halogens or oxygen continues to differ by 2 from the atomic number in B and C, but all ten electrons become inert in Ne. Such considerations, although suggestive, evidently cannot carry us far towards understanding the electronic arrangement.

*The grouping of atomic electrons deduced from spectroscopy**

Atomic electrons may be disturbed by a variety of means: (a) by a sufficient rise of temperature, as in the ordinary flame tests of qualitative analysis (first lines of the principal spectral series usually emitted), (b) by rise of temperature combined with electrical stress, as in the arc or spark, (c) by electron bombardment. The effect of the input of energy occurring in any of these methods may be pictured as a temporary translation of one or more electrons away from the nucleus. On the electron subsequently regaining its normal position, the absorbed energy is emitted as radiation. It is a fundamental postulate of the quantum theory that the frequency ν of this radiation is deter-

* For an account of 'line' spectra (atomic spectra) see Herzberg, *Atomic Spectra and Atomic Structure* (Dover Publications, New York, 1956.

mined by the energy difference between the 'excited' or disturbed state and the normal state afterwards regained. The characteristic relation is $\Delta E = h\nu$, where h is the so-called Planck's constant (p. 30). Just as the energy can only be emitted as a quantum $h\nu$, so it can only be absorbed as a corresponding 'energy-packet' or quantum of definite magnitude. It follows that 'excited' states *are as definite as normal states, and just as specific in the atomic properties.* We therefore more logically conceive what we termed above a 'translation' as a *transition* of the active electron between states. All that can be said is that an electron leaves one state and an electron appears in the other; an attempt to probe the intervening condition, if any, is probably meaningless in terms of modern conceptions. The quantum relation between (observable) frequency and the energy difference between states of the atom makes possible the attempt to ascertain by the methods of spectroscopy the number and succession of such states specific to a particular atom. The states may from our present point of view be also termed *energy-levels.*

In the first attempt, by Bohr in 1913, the states of the atom were pictured as dependent upon the circulation of the active electron in defined orbits of varying eccentricity, each orbit with its associated potential and kinetic energies corresponding to a non-radiating energy-level or 'stationary' state. No explanation was offered of why the states are non-radiative. Motion within the orbit was assumed to be governed by classical laws of dynamics. Although great advances were made by the detailed application of such ideas, Bohr's theory in its original form was found in the end inadequate. A new mechanics applicable to electron motion, termed quantum mechanics or sometimes wave mechanics, has been constructed, which although in theory applicable to the motions of all matter, differs in results from classical mechanics only when very small inertias are involved. A peculiar limitation attaches to this new theory, known as *the principle of indeterminacy*, in that its equations can yield on solution precise values only for a selected number of the factors involved in motion. For example, if, as is the case in experimental spectroscopy, the momentum of the electron is precisely known, then its position can only be expressed as a probability. This new advance has had two consequences: first, a clear explanation of why bonds preserve a mutual direction in space (p. 108), an explanation of the greatest value, and one entirely lacking in the earlier theories; secondly (and this is a troublesome complication to the non-mathematical), the abandonment of the well-defined and simply-shaped orbits of the Bohr theory, and their replacement by difficultly pictured 'electron-clouds'. We shall, however, continue to use the word 'orbit', understood in this wider sense of *the region occupied by the electronic charge*.

The application to the facts of spectroscopy of the principles very briefly alluded to above has shown that electrons may occupy in atoms a considerable variety of orbits, the extent of the variety increasing, as might have been expected, with the number of atomic electrons, i.e. with atomic number. We shall in this chapter continue to limit our attention to the elements H—F in the first series of the periodic classification. Here we shall require to recognize only two types of orbits:

(i) *s*-orbits, (ii) *p*-orbits;

the descriptive letters *s* and *p* are drawn from the close spectroscopic association of certain characteristic spectral series called the *sharp* and the *principal* series, with the transitions of electrons between these *s* and *p* types of orbit (p. 148).

If an atom absorbs energy, at least one of its electrons must undergo transition from its original orbit to a higher energy-level, and therefore to a new orbit. We must therefore conceive that orbits are available to receive an electron at all levels, and we are led to classify orbits (in atoms of the first short series) in two ways, first as *s* or *p*, which gives the type of 'motion', and secondly according to energy-level, or 'shell'. The shells or levels are denoted by the integers, and we therefore speak of $1s$, $2s$, ..., ns orbits, and of $2p$, $3p$, ..., np orbits, in such compact nomenclature condensing both types of information.

All *s*-orbits are *spherically* symmetrical, which description must be carefully distinguished from *circularly* symmetrical: *s*-orbits are therefore unique. On the contrary, in all shells, except the first and lowest, *three* separate but identical *p*-orbits are available, the axes of which are mutually directed at right angles (fig. 28).* In the lowest shell only the 1*s*-orbit is possible. Not infrequently it is convenient to use the phrase 'an *s* (or *p*) electron' in writing on this subject, but if this is not properly understood to mean in full 'an electron occupying an *s* (or *p*) orbit' it may lead to the grave misconception that electrons can in any way be *inherently* distinguished.

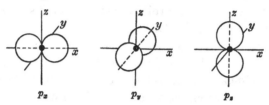

Fig. 28. Representation of *p*-orbits.

Let us now turn to the constitution of the He atom, which contains two electrons. The study of the He (arc) spectrum ordinarily excited in a gas-discharge tube shows indubitably that *both* electrons occupy, in the state of lowest atomic energy or 'ground-state', the 1*s*-orbit, that is, the *same* orbit. Now it is an established principle of modern physics, first announced by Pauli,† that two electrons doing, to write colloquially, *exactly* the same thing, cannot be distinguished by any experimental means. Hence, as it is certain that we *can* discover two electrons in He, and equally certain that they occupy the *same* orbit, some further means of *distinction* between them must exist. When we pass to the next element in order, Li, its normal unit positive charge as an ion, and its spectrum, all demand that in its ground state two of the three electrons are, as in He, in the 1*s*-orbit, tightly held to the nucleus, and the third more loosely held in a 2*s*-orbit. Under no conditions do all three electrons come to occupy the 1*s*-orbit. It is therefore clear that for two electrons occupying the 1*s*-orbit the necessary distinguishing feature is to

* *d*-Orbits (from the 'diffuse' spectral series) are *five* in number, each of which is differently directed in space. These orbits have importance in spectroscopy, but none in the chemistry of the elements H—Ne.

† For a more exact statement of the principle see p. 148.

be so chosen that it distinguishes two things but cannot be used to distinguish three. The feature chosen is *spin*, which in being conceivable only in two opposed directions, clockwise and anti-clockwise, admirably fits the requirements. It is doubtful if any simple picture of electron spin should be encouraged, even if it were possible to make such a picture out of electron-clouds. The genuineness of the spin motion is, however, confirmed by the known fact that the spins of electrons can couple so as to neutralize each other's effects (but not of course to abolish either spin), and that, if this coupling is not complete in an atom or molecule, that atom or molecule exhibits *paramagnetism* (cf. II, p. 26 and V, p. 159). The fact that so few atomic and molecular structures are found to be paramagnetic argues for a high tendency towards this sort of coupling (p. 122).

The question arises—what changes occur in the helium atom if by any cause both electrons acquire spins in the same direction ('parallel' spins)? They cannot now both occupy the same $1s$-orbit, and hence one must pass to a $2s$-orbit, at a higher energy-level. Thus for anti-parallel spins the ground-state is $1s^2$, i.e. two electrons both in the $1s$-orbit; for parallel spins it is $1s\,2s$, the energy of the electron in the $2s$-orbit being approximately equal to that of the first *excited* level of the anti-parallel configuration. The spectral lines associated with the $1s^2$ configuration will be due to electron transitions between levels having that of the $1s$-orbit as base; those associated with the $1s\,2s$ configuration will be due to transitions of the electron in the $2s$-orbit, and will have this much higher level as base. The two sets of spectral lines in fact reveal themselves as completely different spectra, apparently produced by one and the same element (fig. 32, p. 144). These expectations correspond in every detail with long-established experimental facts about the spectra of helium. So struck were spectroscopists by the apparent anomaly of a single element yielding two distinct spectra, that the helium atom was thought to exhibit a sort of atomic 'allotropy', the forms being named *orthohelium* (ground-state $1s\,2s$), and *parahelium* (ground-state $1s^2$); only the development of the latest ideas caused the abandonment of this assumption (in 1925). The subject of the electronic configuration in helium has been explained in some detail, because, as will soon appear, it embodies the main ideas in the spin theory of chemical union. One quite obvious extension of Pauli's principle may at once be stated. *The maximum number of electrons that can be accommodated in any one orbit is two.* Hence in all shells the *s*-orbits and each of the three *p*-orbits may all contain not more than two electrons each; there will thus never be more than six '*p*-electrons' or more than two '*s*-electrons' in any shell. With this principle in mind as a guide in distribution, we may now proceed to form the normal (ground-state) configuration of the first-row elements H to Ne (Table 19).

We note that (*a*) all elements after hydrogen contain a completed 1*s* shell, (*b*) after lithium all elements, in addition to the completed 1*s* shell, contain two electrons in the (completed) 2*s* level, (*c*) Ne has both shells 1 and 2 completed, eight electrons in all being in the shell 2. Further, it will be noted that all pairs of '*s*-electrons' *must* be coupled anti-parallel; while a maximum of three *p*-electrons, as in nitrogen, if they are in *different p*-orbits, are not to be regarded as coupled in the same sense. The directions of the spins of two electrons necessarily paired in the *same* orbit are called '*restricted*': the spins of 'lone' electrons are to be regarded as '*unrestricted*', *even though the spins of two of the 'lone' electrons may be anti-parallel.* If there are four *p*-electrons, as in oxygen, two of them must be in the same orbit and therefore coupled anti-parallel and both restricted. In fluorine only one electron remains unrestricted, and in neon none. Looking at the column in Table 19 headed 'Chemical valency (to hydrogen)' we see that zero valency is associated, in helium and neon, with the complete absence of unrestricted electron spin. It is a reasonable deduction that valency depends in some way on the number of electrons with unrestricted spin. Inspection of the table shows in fact that the number of *unrestricted* electrons in fluorine, oxygen, nitrogen, lithium and hydrogen equals the chemical valency, but this simplest of relations fails for the remaining elements beryllium, boron and carbon. For F, O, N, Li and H, we notice immediately that if all the unrestricted electrons in one of the atoms occupied the vacant orbits of a second atom, as well

Table 19. *Electronic configurations of atoms H to Ne*

	Atomic no. Z	$1s$	$2s$	$2p$			Configuration in full	Unrestricted electrons	Chemical valency (to hydrogen)
H	1	↑					$1s^1$	1	1
He	2	↑↓					$1s^2$	0	0
Li	3	↑↓	↑				$1s^2 2s$	1	1
Be	4	↑↓	↑↓				$1s^2 2s^2$	0	2
B	5	↑↓	↑↓	↑			$1s^2 2s^2 p$	1	(?)
C	6	↑↓	↑↓	↑	↑		$1s^2 2s^2 p^2$	2	4
N	7	↑↓	↑↓	↑	↑	↑	$1s^2 2s^2 p^3$	3	3
O	8	↑↓	↑↓	↑↓	↑	↑	$1s^2 2s^2 p^4$	2	2
F	9	↑↓	↑↓	↑↓	↑↓	↑	$1s^2 2s^2 p^5$	1	1
Ne	10	↑↓	↑↓	↑↓	↑↓	↑↓	$1s^2 2s^2 p^6$	0	0

as their own original orbits, i.e. if two electrons were 'shared' for fluorine and hydrogen, four for oxygen, and six for nitrogen, each atom of the pair F_2, H_2, O_2 and N_2 would reach the condition of neon, with *probably* a large increase of stability over the original single atoms (fig. 29). Hence we could predict the formulae H—H, F—F, O=O, N≡N, wherein each bond-line corresponds to a *pair of electrons necessarily coupled anti-parallel*, originating

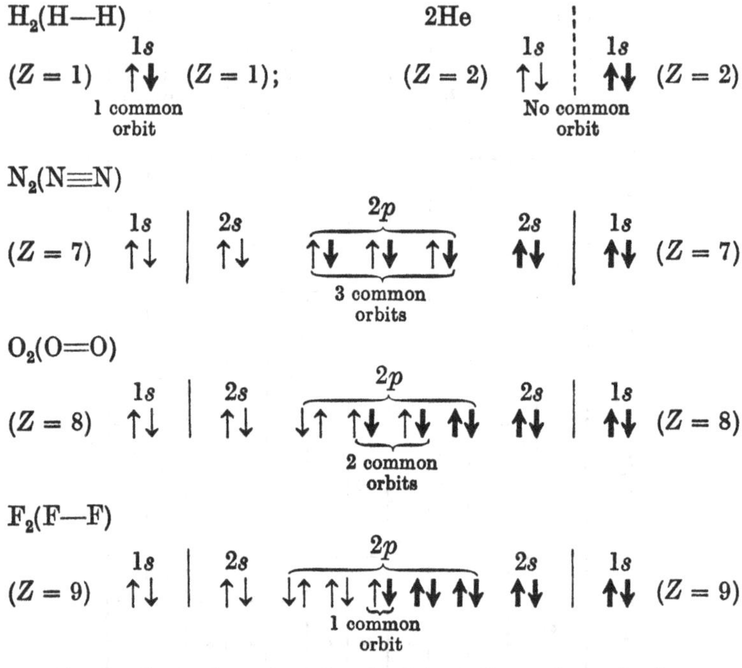

H_2(H—H)
$1s$
$(Z = 1)$ ↑↓ $(Z = 1)$;
1 common orbit

2He
$1s$ $1s$
$(Z = 2)$ ↑↓   ↑↓ $(Z = 2)$
No common orbit

N_2(N≡N)
$1s$ $2s$ $2p$ $2s$ $1s$
$(Z = 7)$ ↑↓ | ↑↓ | ↑↓ ↑↓ ↑↓ | ↑↓ | ↑↓ $(Z = 7)$
3 common orbits

O_2(O=O)
$1s$ $2s$ $2p$ $2s$ $1s$
$(Z = 8)$ ↑↓ | ↑↓ | ↓↑ ↑↓ ↑↓ ↑↓ | ↑↓ | ↑↓ $(Z = 8)$
2 common orbits

F_2(F—F)
$1s$ $2s$ $2p$ $2s$ $1s$
$(Z = 9)$ ↑↓ | ↑↓ | ↓↑ ↑↓ ↑↓ ↑↓ ↑↓ | ↑↓ | ↑↓ $(Z = 9)$
1 common orbit

Fig. 29. Electronic configurations for diatomic (homopolar) molecules, according to the spin theory of valency.

equally from each atom of the compound molecule, and represents the so-called *covalent bond*. In this conception we advance in one step to an explanation of what has been earlier referred to (II, p. 19) as the most elusive character of chemical combination. By a simple extension of the same idea we see how the hydrides FH, OH_2 and NH_3 come to be formed. Fig. 30, for example, portrays the combination of nitrogen and hydrogen to form ammonia. The s-orbit in the H atom and the three p-orbits in the N atom each hold only one electron before combination takes place. This event may be pictured as resulting from a spatial *overlapping* of a *p*-orbit in the N atom by an *s*-orbit in an H atom, so that both orbits become common to each atom, and both are fully, i.e. doubly, occupied (see further, Chapter VI).

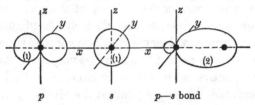

$$p \qquad\qquad s \qquad\qquad p\text{—}s \text{ bond}$$

Fig. 30. Representation of overlapping of singly-occupied *p*-orbit (e.g. in nitrogen) with singly-occupied *s*-orbit (e.g. in hydrogen) to form a doubly-occupied bonding orbit.

We have been careful to write above that there will *probably* be a large gain of stability, basing this notion on the qualitative analogy with the entirely stable atoms of helium and neon. This argument cannot be accepted as completely convincing until it is shown that in fact the energy change equals the experimental value given in the equations

$$2H = H_2 + 103 \cdot 4 \text{ Cal.} \qquad\qquad 2N = N_2 + 225 \text{ Cal.}*$$
$$2O = O_2 + 117 \cdot 4 \text{ Cal.} \qquad\qquad 2F = F_2 + 38 \text{ Cal.}\dagger$$

For hydrogen complete calculations have been carried through and lead to a result in almost exact agreement with the experimental value of the heat of dissociation; for oxygen, however, such calculations, if effected, could hardly lead to such good agreement, for our model fails to explain why alone among these diatomic molecules oxygen has strong magnetic properties. Moreover, while it is well adapted to explain the formation of the peroxide ion $O_2^=$, which is the electronic analogue of F_2, nothing can be learnt from it about the superoxide ion O_2^-. A more comprehensive treatment of O_2 follows later (p. 191). We owe the first qualitative suggestion of the electron pair as the unit of the bond to the remarkable intuition of G. N. Lewis, although Kossel had already initiated the ionic approach, afterwards developed

* Douglas, *J. Physical Chem.* 59, 109 (1955).
† Sharpe, *Quart. Rev.* 11, 49 (1957).

by Fajans (p. 117). At this time (1916) the concept of electron spin and its significance had still to be formed,* and the rise and elaboration of quantum mechanics was needed before Heitler and London† could give the first quantitative treatment of the simplest reaction $2H = H_2$.

We must now consider the valency of those elements, beryllium, boron and carbon, whose electronic configuration only admits of a maximum valency two units lower than the established chemical values. In respect to carbon we may recall that a school of organic chemists, with which the name of J. U. Nef is most closely associated, for long supported the bivalency of carbon in certain of its compounds, notably in acetylene, carbon monoxide, the isonitriles, and in the fulminates (derivatives of fulminic acid HONC). If we accept the electronic scheme for carbon shown in Table 19 we must find carbon bivalent in *all* organic compounds, a situation quite caricaturing the position of the school of Nef, whose opinions have as a matter of fact fallen into disrepute in recent years. On the other hand, the spectroscopic evidence puts beyond all doubt the electronic configurations of beryllium, boron and carbon indicated in Table 19 (the ground-states are ^1S, ^2P and ^3P for Be, B and C respectively, see v, p. 148). Adopting the principles we have already discussed that seem to work well for the elements other than those now being considered, we require four 'unrestricted' electrons in carbon to confer quadrivalency, three in boron, and two in beryllium. Some process of 'uncoupling' one of the $2s$-electrons and placing it in a p-orbit must therefore be contemplated. Such a 'promotion' will require the input of considerable energy (about 100 Cal. g. atom),‡ but recalling that the energy released as heat when an *atom* of carbon combines with four hydrogen atoms to yield methane is 394 Cal. (p. 115), it would seem that an adequate source of 'promotion' energy is available.

The mere conversion of the carbon configuration s^2p^2 into sp^3 will, however, not suffice, for we are certain that the four bonds in CH_4 (or in CCl_4) are identical in nature, and mutually

---

* Goudsmit and Uhlenbeck, *Naturwiss.* 13, 953 (1925).
† *Z. Physik*, 44, 455 (1927).
‡ van Vleck, *J. Chem. Physics*, 2, 20, 297 (1934).

directed tetrahedrally. Now not only is the s-orbit essentially different in descriptive equations from p-orbits, but its spherical symmetry (p. 107) precludes it from being associated with a particular spatial direction. It is therefore necessary to weld, or 'hybridize', the orbits sp^3 into four equivalent orbits. This mathematical operation shows that four such orbits can be formed stably (fig. 31), but also offers the interesting alternative solution of three equivalent orbits directed to the corners of an equilateral triangle, and a fourth directed at right angles to the triangular plane and weaker in bonding

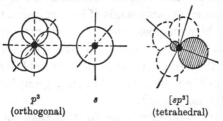

$$p^3 \qquad s \qquad [sp^3]$$
(orthogonal)   (tetrahedral)

Fig. 31. Hybridization of orbits in the carbon atom.

power than the others. This possibility will be further considered in relation to the structure of ethylene and benzene (p. 194). It must be confessed that the necessity of manipulating the configuration of the carbon atom in order to ensure the correct valency may justifiably leave an impression of special pleading; it is therefore satisfactory to find that the calculated heat of formation of methane agrees excellently with the experimental figure 394 Cal.* Having once accepted this treatment for carbon we readily agree to the change s^2p to $[sp^2]$ for boron, and s^2 to $[sp]$ for beryllium (the sign [ ] means hybridization). For Be, B and C the ground-state and the state of maximum valency are therefore not identical. It is to be noted that $[sp^2]$ gives 3 *coplanar* bonds (as in BF_3, p. 59); $[sp]$ 2 *collinear* bonds (as in $BeCl_2$, p. 56).

Returning for the moment to nitrogen, oxygen and fluorine we see that in these elements promotion of one electron would not

* Coulson, *Trans. Faraday Soc.* 33, 388 (1937).

increase the number of 'unrestricted' electrons (and therefore not the valency), for in these atoms there is at least one occupant in all possible orbits in the ground-state. Hence we may at once assign maximum possible valencies as follows: (N) = 3, (O) = 2, (F) = 1. The inertness of helium would be removed if one of its (s) electrons were promoted to a higher ($2s$) level. From the spectrum of helium (p. 145) it is known that the energy required to effect this change is some 460 Cals./g. atom. Such intense localizations of energy cannot occur in chemical reactions, but of course are normally reached in discharge tubes emitting the helium spectrum. Under these conditions there may be some chance of discovering compounds of helium (see p. 183), but their exceptionally high intrinsic energy would very probably limit their lives to extremely short periods. The promotion of a $1s$-electron to the second shell is no more probable for other atoms, in *chemical* reaction: hence the 'valency group' of electrons is confined to the outer shell.

It has already been mentioned (p. 105) that numerous modes of exciting spectra are in use. One or more electrons in a *valency* group, when excited to higher levels, emit on return to their normal orbits 'lines' in the *visible spectrum* ($\lambda = 7\cdot4 - 4\cdot0 \times 10^{-5}$ cm.). If the excitation is only by a moderate rise of temperature, as in the flame-tests of qualitative analysis, one electron of the valency group is excited only to the next higher levels, and on return emits the first (lowest frequency) lines of the 'principal' series (e.g. D lines of sodium). If an exceptionally powerful mode of excitation is employed, for example a beam of high-speed electrons, disturbance may penetrate deep into the atom and excite one of the two innermost $1s$-electrons. Such an electron, in regaining its ground-state by a transition in the intense field of the net positive charge $Z - 1$, emits radiation of very high frequency, usually called *X-radiation* ($\lambda \sim 10^{-8}$ cm.). Moseley* demonstrated that the frequency ν_K of this radiation is related to the positive charge $Z - 1$, and therefore to Z, by the simple law

$$\nu_K = \text{const.}\,(Z - 1)^2.$$

The lines in the X-ray spectrum so emitted are usually called

* *Phil. Mag.* (6), 26, 1024 (1913); *ibid.* (6), **27,** 703 (1914).

the K lines, and the $1s$-electrons are therefore often referred to as K electrons. Hydrogen has too small a nuclear charge for the radiation emitted by it under any mode of excitation to reach beyond the ultraviolet region (the Lyman series; $\lambda \sim 2 \times 10^{-5}$ cm.). Further, the scattering effect of combined hydrogen upon X-radiation is relatively negligible. This means that the methods of investigating crystal structure based upon the diffraction of X-radiation (III, p. 38) cannot be used to locate hydrogen atoms—a serious limitation, particularly in examining organic compounds.

The problem of the nature of the chemical bond may be approached from another angle (Fajans). Attention has been drawn (II, p. 26) to the significant fact that each inert gas is preceded in the periodic table by a halogen, and succeeded by an alkali metal. If one atom of the metal, say Li, released one electron *completely* to a halogen, say F, the resulting ions Li^+ and F^- would resemble in electronic configuration (but not otherwise) an inert gas (Ne). Hence we explain the vigorous reaction

$$Li + F = Li^+ . F^-, \quad \Delta H = -146 \text{ Cal.}$$

The product, an ion-pair, is a typical electrolyte, and the two ions exist independently in all states—in the solid crystal and in aqueous solution. No electrons are shared, and there is no *covalent* bond (p. 112) between the ions, which however experience a strong mutual electrostatic attraction. We might imagine other salts, such as the halides of copper and silver, to arise in a similar way, and, according to the crystallographic evidence (p. 53), such a view would be justified for AgCl and AgBr. Although in the solid solutions of AgI in AgBr silver and iodine ions exist stably as separate units, in pure AgI at normal temperature no ions of either sign are present: the salt is a macromolecular structure composed of Ag and I atoms in *covalent* union. Both cupric chloride and bromide are freely soluble in water, and yield in dilute solution (hydrated) ions Cu^{++}, but neither solid contains any such ions, both solids being again structures in which metal atoms and halogens are co-valently linked (p. 53). Cuprous chloride, bromide and iodide all simulate AgI in macromolecular solid structure. The examples of the cupric halides (and of many other salts) show that while macromolecular solid constitution can hardly favour solubility in water, it does not necessarily prevent it. We may suppose that when Ag^+ (from $AgNO_3$) and I^- (from KI) are brought into contact the positive ion is able to draw one or more electrons away from the (large) anion, in which the outer electrons are loosely held, while simultaneously repelling the halogen nucleus. In short, the anion undergoes a marked distortion, which may reach a point, as in AgI, where the electrons withdrawn become common property of both atoms. Such deforming actions of cations upon anions will be most drastic (a) when the ionic charges are high, (b) when the anion is large and the cation small, (c) when one or both of the ions does *not* resemble an inert gas in electronic configuration.

These views are exceedingly useful in co-ordinating many otherwise puzzling facts about metallic compounds, and especially in elucidating cases of compounds formally to be regarded as salts, but which lack the properties of electrolytes. Besides the salts above, other typical examples are the salts of mercury (in both series) and $SnCl_4$. $(Sn . 2H_2O) Cl_2$ is a true electrolyte, but the great deforming power of the (smaller) stannic ion Sn^{4+} so distorts the halogen ions that $SnCl_4$ closely resembles CCl_4 (cf. p. 71). It is to be expected that the ions Cu^+ and Cu^{++} would both be smaller than Ag^+, and in accordance with (b) above the smaller cations interact drastically with all the anions Cl^-, Br^- and I^- ($r = 1·81$, $1·95$, and $2·16$ respectively), while Ag^+ so interacts only with the largest halide ion. Also the existence of a molecular *polarity*, depending inversely on the extent of the ionic deformation, is readily predictable (p. 120). In spite of these practical advantages it cannot be said that this viewpoint shows the clear-cut picture of the chemical bond and the reasons for its formation gained by the first method. For the bond formed by two paired and shared electrons (i.e. with anti-parallel spins, and occupying one orbit in each atom) we shall use the now well-known term '*covalent bond*'; for the purely electrostatic linkage between *ions* we shall use the term '*ionic bond*'.

*Outline of the mathematical technique for calculating
electronic distribution in a molecule AB*

(1) The nuclear charges (atomic numbers), and the inter-nuclear distance are assumed known and fixed (see methods (3) and (4), III, p. 38 sqq.).

(2) A wave function (or orbital) for two electrons (1, 2) *equally* shared in the bond A—B is derived from those of the separate atoms, ψ_A and ψ_B, and takes the form

$$\psi_{cov} = \psi_A(1)\,\psi_B(2) + \psi_B(1)\,\psi_A(2).$$

(3) Corresponding functions for the ionic bonds A^+B^- and A^-B^+ are $\psi_B(1)\,\psi_B(2)$ and $\psi_A(1)\,\psi_A(2)$ respectively.

(4) The appropriate function for AB is then assumed to be

$$\Psi = a_0\psi_{cov} + a_1\psi_{B^-} + a_2\psi_{A^-}.$$

(5) Values of the coefficients a are now deduced, which give the compound *minimum energy* (see further, p. 188).

The method of combining the sets of expressions (2) and (3) resembles the classical problem of combining the motions of vibrating systems, such as pendulums, and involves the principle of *resonance*. The technique of combination is commonly described as 'hybridization' and the final single structure deduced from the process as a 'resonance hybrid'. The welding of the *s*- and *p*-orbits in the carbon atom after promotion is another example of hybridization (p. 115). When the molecule of an elementary substance such as H_2 is being treated, it is obvious from mere considerations of symmetry that the bond A—A is non-polar, and that the (ionic) coefficients a_1 and a_2 will be equal, and the two corresponding ionic forms self-neutralizing. We should, however, not be justified in neglecting the ionic functions, i.e. in setting $a_1 = a_2 = 0$, for it is found that *every constituent of the hybridization contributes to lower the energy of the final hybrid*, and so increases its stability.* The additional liberation of energy of formation so brought about is known as 'resonance energy', and its discovery is peculiar to modern methods. The most recent calculations of the energy of forma-

* See the discussion of CO_2, p. 129; also, p. 208.

tion of H_2 show that the ionic forms contribute only about 5 per cent. For the hydrogen halides the ionic contributions decrease, as would be expected, as the atomic number of the halogen rises: the following approximate values may be quoted*

HF, 43; HCl, 17; HBr, 13; HI, 5 per cent.

Here, except possibly in HI, the only important ionic form is H^+X^-. In the general case the coefficients a_1 and a_2 are unequal, and the normal covalent bond A—B is found to be *polar*, in harmony with experimental results on electric moments† (III, p. 38).

It is of the first importance to appreciate that there are *not* two separable molecules, e.g. A—B and the ion-pair $A^+.B^-$, in a sort of tautomeric equilibrium. *There is only one actual molecule*, but the mathematical difficulties of a more direct approach to ascertaining the configuration of the bonding electrons leading to minimum energy are at present very great. Direct methods, however, have been steadily developed, especially under the leadership of Hund, Mulliken, Herzberg and Lennard-Jones. We shall in subsequent discussions of section II (based on the above methods) be constantly required to assume this idea of resonance hybrids, but it must at the outset be realized that the 'hybridization' is forced upon us, not because the unit structures of the hybridization have any self-existence, but by present shortcomings in mathematical technique (and pictorial representation). It must be specially noted that the condition of section (1) above limits resonance to structures *with identical nuclear framework* (see p. 127).

To the non-mathematical the conception of resonance hybrids is a difficult one, especially as there is no obvious manner of pictorial representation. A very simple but crude analogy may prove helpful. Suppose a printer is asked to produce a handbill in green colour, at short notice. He discovers to his consternation

* Wells, *Structural Inorganic Chemistry*, Oxford, 2nd ed. (1950), p. 37.

† Inequality between a_1 and a_2 is not directly related, as might have been thought, to the difference in *atomic numbers* Z_A and Z_B, but to the difference in *electro-negativity* between the atoms A and B (see further, pp. 224 and Mulliken, *J. Chem. Physics*, 2, 782 (1934)).

that he has exhausted his green ink, but is well supplied with blue and yellow. He therefore carefully over-prints a blue bill with the yellow ink, and will, if his 'registration' is nice enough, not raise any doubts in his customer that the normal printing in green ink has been carried out. The reader will see here the blue and yellow 'units' combining to give a *single* green 'hybrid'. In more advanced colour-printing more colour 'units' could be used, but there would in all cases be only *one* final picture, and those ignorant of colour-printing methods could not deduce the presence of more than one impression.

SECTION II

The bond-diagrams of some simple molecules, radicals and ions formed from first-row elements (H to F)

In arriving at the constitution of a molecule, all legitimate arrangements of the available (valency) electrons on the nuclear framework must be considered (such 'unit structures' are often termed 'canonical forms'). For first-row elements arrangements which infringe the 'octet rule', i.e. exhibit an atom containing more than eight electrons in the second shell, are to be excluded. When the full number of possible arrangements has been set out, a process of hybridization (see above) yields the final *singular* molecular structure. In devising electronic arrangements it is not necessary, nor usually convenient, that the scheme of building up the constituent atoms into the finished structure should correspond with actual chemical reactions, but all such alternative constructions must be examined in order to secure a complete set of canonical forms, and the correct energy of formation (p. 119). Thus we may construct N_2O in three ways: (*a*) by combining the atoms N, N, O in this order simultaneously, (*b*) by combining O with N_2, (*c*) by combining N with NO. None of these processes corresponds to a normal chemical change, nor bears any obvious relation to the methods of preparing nitrous oxide. In bond-diagrams we shall usually represent shared (bonding) electrons by the long-familiar stroke; when necessary unshared ('lone') electron pairs will be indicated by the colon sign, thus

$$:N\equiv N: \quad \text{and} \quad ::O{\displaystyle{\overset{\displaystyle H}{\underset{\displaystyle H}{<}}}} ,$$

but this symbolization is just an *aide-mémoire*, and must not be interpreted as indicating a *fixed locality* for the electrons. It will be convenient to represent (resultant) polarity by signs *below* the atoms concerned, thus A_+—B_-. Signs placed *above* the atoms, e.g. A^+—B^-, will indicate a co-ionic bond (see p. 125). Frequent use will be made of the empirical *isoelectronic principle* 'simple structures containing the same number of valency electrons are represented by the same bond-diagram'.* To facilitate the application of this principle molecular structures have been classified in Table 20 according to their content of valency electrons.

Table 20. *Isoelectronic structures*

Total valency electrons	Examples from first-row elements
8	Beryllium oxide, boron nitride, diamond, (C_2), silicon carbide: CH_4, NH_3, H_2O, HF; $(NH_4)^+$, $(H_3O)^+$
10	N_2, CO, C_2H_2; —CN, CN^-, —NC, NO^+
11	NO
12	O_2, H_2CO, C_2H_4; —NO
13	O_2^-
14	F_2; $O_2^=$ (as in BaO_2), H_2O_2, HO_2^-; (ClO^-, BrO^-, IO^-)
16	CO_2, N_2O; —CNO, —NCO, —ONC, —N_3; CH_2N_2, NO_2^+
17	NO_2
18	$NOF(NOCl)$, $(CN)_2$, O_3, —NO_2, NO_2^-, HCO_2^-
24	$BO_3^=$, $CO_3^=$, NO_3^-, BF_3
26	NF_3

The tendency of electrons to occur in compounds with coupled spins is emphasized by the rarity of structures with an odd number of electrons. The presence of uncoupled spin in NO, NO_2 and O_2^- is manifested by their property of paramagnetism.

In order to bring out certain important novel features it will be convenient to deal first with the 10-electron group.

* Penney and Sutherland, *Proc. Roy. Soc.* A, **156**, 654 (1936).

*The structures —CN, —NC, CO and NO⁺, and
the fulminate radical —ONC*

The radicals —CN and —NC are well known to organic
chemists as the nitrile and isonitrile groups. The isonitrile group
has been held to contain bivalent carbon (p. 69). We must
remember that, when these radicals are in combination with
another radical R, giving the compounds RCN and RNC, the
radical R supplies one electron to the bond. The available
valency electrons may thus be set out:

—C N   —N C   C O $:N\equiv N:$ $N\equiv O^+$
1+4   5   1+5   4   4   6   5   5   5   5

The structure of N_2 and NO^+ follows from symmetry. Using the
isoelectronic principle we write —CN immediately as

$$:C_+\equiv N_-:$$

the polarity ($\mu = 3.6$) being introduced (Table 4, p. 38). In
the two other groupings the distribution of electrons is unsymmetrical; we now convert into the symmetrical form of
nitrogen and —CN by transferring an electron from O or N respectively to C, thus forming the hypothetical ionic groupings

—N⁺ C⁻ and C⁻ O⁺
5   5   5   5

These ions are finally united, again in an application of the isoelectronic principle, thus

$$:N^+\equiv C^-: \text{ and } :C^-\equiv O^+:$$

It is to be noted that the input of energy required to transfer
an electron from O or N to C is almost wholly regained in electrostatic interaction as the ions approach each other and combine (see also p. 135).

In support of the detailed similarity in the diagrams of CO and
N_2 it may be mentioned that the van der Waals' constants a and
b are nearly equal for the two gases and consequently they have
nearly the same boiling points, critical points, etc.

We meet for the first time in the structures of —NC and CO a
superposition of covalent and ionic bond, and we can immediately
understand the unexpected lack of polarity in carbon monoxide

$(\mu < 0.1)$. The grouping C≡O would, to judge from the polarity of C=O (Table 4), be exceptionally polar, with the negative charge on the *oxygen* atom. The addition of the ionic charges yields a strong neutralizing dipole in the opposite direction. In —NC the polarity of the triple bond is less than in C≡O, and therefore the polarity of the ionic displacement decides the polarity of the structure and we have a net polarity thus

$$—N_+ ≡ C_-$$

It has been mentioned before (p. 68) that some chemists were willing to assume quadrivalent oxygen in CO, and others bivalent carbon. What is the valency of these atoms in the above structure? If the positions of all the electrons are shown in detail it is seen that both the carbon and oxygen have normal complete octets, as they have in CH_4 and H_2O respectively. It is reasonable therefore to assert that each atom is exerting its usual valency in CO, in spite of an apparent contrary appearance in the bond-diagram. We reach here a new stage in valency problems, in that we can no longer assume that valency is immediately deducible by counting the bonds issuing from an atom (p. 16).

In the structure $: C^- ::: O^+ :$, the six bonding electrons are not supplied equally by the two linked atoms; *two* arise from the carbon, and *four* from the oxygen; in other words, one pair has arisen from the oxygen *alone*. It is because of the unilateral nature of this electronic transaction that exerted valencies are not immediately obvious from an inspection of bond-diagrams containing bonding pairs of this kind (i.e. co-ionic bonds, see below). In forming carbon monoxide the oxygen may be said to 'donate' two electrons and the carbon to 'accept' them, the resulting bond being styled a 'dative covalency'. An arrowhead is often attached to the bond to indicate the direction of the donation: $: C \xleftarrow{} O :$

Since bonds formed in this manner play a large part in compounds, named 'co-ordination compounds' by Werner (p. 168), another name in frequent use is 'co-ordinate bond'. Both of these terms, as well as the arrow sign, are unfortunate in emphasizing the mode of formation of the bond at the expense of describing it when formed. In what follows the term *co-ionic bond* will be used, with the intention of stressing its nature as a superposition of a covalent on an ionic bond. It may be noted that the tendency for the oxygen atom to recover the donated pair of electrons causes carbon monoxide to show an unsaturation confined to the carbon atom, as exemplified in the following typical reactions:

$$CO + OH^- = \begin{matrix} O^- \\ \diagdown \\ H \diagup \end{matrix} C{=}O \ (\text{formate}); \ CO + OCH_3^- = \begin{matrix} O^- \\ \diagdown \\ CH_3 \diagup \end{matrix} C{=}O \ (\text{acetate})$$

$$CO + Cl_2 = \begin{matrix} Cl \\ \diagdown \\ Cl \diagup \end{matrix} C{=}O; \ CO + H_2 = \begin{matrix} H \\ \diagdown \\ H \diagup \end{matrix} C{=}O$$

The position of R— in RCN and RNC fixes the initial electronic distribution as $\underset{1+4\ \ 5}{C\ N}$ and $\underset{1+5\ \ 4}{N\ C}$ respectively, and therefore ensures the ordinary *isomerism* of RCN and RNC, the nitriles and isonitriles being chemically independent. Hydrocyanic acid and the cyanides, however, give rise to the ion $(CN)^-$, which, when constructed from either of the two possible groupings $\underset{6\ \ \ 4}{N^-\ C}$ or $\underset{5\ \ \ 5}{N\ C^-}$, assumes the *same* electronic distribution, viz. $N{\equiv}C^-$. In unionized hydrocyanic acid, however, we can have forms HCN and HNC corresponding exactly with RCN and RNC. The fact of ionization brings these two forms of acid into *tautomeric* equilibrium.* There being only one K^+ and only one CN^-, there can only be one form of the ion-pair $K^+ . CN^-$. In silver cyanide, as in silver iodide (p. 53), the metal is probably at least partially covalently linked, and we could therefore expect the existence of both AgCN and AgNC. The crystal structure of silver cyanide fulfils this expectation (see p. 55). The fulminate radical —ONC follows from that of —NC, as the latter may be united either with (univalent) R or with (univalent) RO, giving in the latter case the fulminate as $R{-}O{-}N^+{\equiv}C^-$. The methods of organic chemistry can often be employed to compare directions of polarity as follows. X-ray examination of hexamethyl benzene $(CH_3)_6C_6$, (m.p. 164°), shows that the molecule contains two

* See, however, Herzberg, *J. Chem. Physics*, **8**, 847 (1940).

concentric regular hexagons of C atoms, and therefore we may fairly assume that in the parent substance benzene the bonds C—H are symmetrically directed. It follows that in the p-compound Y—C⟨ ⟩C—X the polarity of C—X will diminish or enhance that of Y—C, according as the polarities of C—X and Y—C are opposed or not. The following data have been obtained:

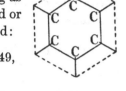

$$\text{Cl}_-\text{—C}_+\!\langle\;\rangle \quad \mu = 1\cdot 55, \qquad \langle\;\rangle\text{—NC} \quad \mu = 3\cdot 49,$$

$$\text{Cl—}\langle\;\rangle\text{—NC} \quad \mu = 2\cdot 07.$$

The fact that the terminal C—N distances in CH_3CN and CH_3NC are equal ($1\cdot 16$ A.) confirms the identity of the bonding.*

The 14-electron group

Although chlorine, bromine and iodine are not first-row elements, it is convenient to discuss the hypo halogen oxy-acids at this point. It will be provisionally assumed that all the halogens have analogous valency groups, similar to that of fluorine. This assumption will be justified later in Chapter v. For —OCl we have —O Cl exactly like Cl Cl. Hence we write
                        1+6 7   7 7
—O—Cl. For —ClO we have —Cl O, and by the process used
                                1+7 6
before, we find for this —Cl$^+$—O$^-$, i.e. a single co-ionic bond. The *ion*, like (CN)$^-$, can have only *one* structure, Cl—O$^-$, formed from either of the groupings Cl$^-$ O or Cl O$^-$; Thus hypochlorous
                                       8   6   7 7
acid, like hydrocyanic acid, could exist in two *tautomeric* forms, HClO and HOCl. There is, however, no generally accepted evidence for such a condition, and hypochlorous acid is supposed to consist of the single species HOCl. On the other hand, the production of chlorhydrins, by the addition of hypochlorous acid to the ethylenic bond (p. 90), as well as the interaction of hypoiodous acid with HCl (p. 93) would be neatly explained by postulating for these acids not only the normal acid form of

* Brockway, *J. Amer. Chem. Soc.* **58**, 2516 (1936).

ionization, but in appropriate conditions, also a basic form $HOX = OH^- + X^+$.

H_2O_2, and its ion $(HO_2)^-$, also belong to the 14-electron group. For H_2O_2 there are two possible but non-resonating structures

$$\begin{array}{c} H \\ \diagdown \\ \underset{8}{O} \!-\! \underset{8}{O} \diagup^{H} \end{array} \quad \text{and} \quad \begin{array}{c} H \\ \diagdown \\ H \diagup \underset{8}{O^+} \!-\! \underset{8}{O^-} \end{array}$$

(cf. p. 82). For the ion we have only $H\!-\!O\!-\!O^-$.

The above examples—*isomerism* of the nitriles and isonitriles, *tautomerism* in hydrocyanic acid and in hydrogen peroxide—enforce the principle that resonance (and hybridization) is only possible between structures *with identical or nearly identical nuclear framework* (see further VI, p. 213).

The 'odd electron' molecules NO, NO_2 and O_2^-

In these molecules completed octets are impossible. If for NO we write $\cdot\underset{7}{\overset{\cdot\cdot}{N}}\!=\!\underset{8}{O}$, the bond-diagram shows a 'lone' (unpaired) electron, confirmed by the paramagnetism of the gas (p. 160). This molecule is of the nature of a free radical, such as CH_3, but it is remarkable that NO shows no tendency to polymerize.

We form NO_2 by uniting NO with O, as in the actual chemical reaction:

$$\underset{8}{O}\!=\!\underset{7}{\overset{\cdot\cdot}{N}}\!\cdot \quad \underset{6}{O} \;\rightarrow\; \underset{8}{O}\!=\!\underset{6}{\overset{\cdot\cdot}{N}}{}^{+} \quad \underset{7}{O^-} \;\rightarrow\; \underset{8}{O}\!=\!\underset{7}{\overset{\cdot}{N}}{}^{+}\!-\!\underset{8}{O^-}$$

NO_2, like NO, is paramagnetic, and, as is well known, dimerizes readily to

$$\begin{array}{c} O\!=\!N^+\!-\!O^- \\ \overset{\cdot\cdot}{} \\ O\!=\!N_+\!-\!O_- \end{array}$$

It is triangular in shape: $\underset{O\diagup}{\overset{N^+}{}}\diagdown_{O_-}$

Two structures closely related to NO_2 are the nitro-group $-NO_2$, and the nitrite ion NO_2^-. In these the extra electron from the bond or from the ionic charge completes the octet of the nitrogen, and we have

$$-\underset{8}{N^+}\!\!\diagup_{\diagdown O_-}^{O} \quad \text{and} \quad {}_{O}\!\diagup^{\underset{8}{N}}\!\diagdown_{O_-}$$

In the nitroso-group —NO, also, the bond confers the necessary extra electron to complete the nitrogen octet. It is noteworthy and in accordance with expectation that the nitro-group, the nitroso-group and the nitrite ion are all *diamagnetic*. In all NO_2 structures resonance between congruent forms ($O{=}N{-}O$ and $O{-}N{=}O$) operates to equalize both N, O bonds, a fact elegantly demonstrated by the non-polarity of p-dinitrobenzene (cf. p. 126).

Nitric oxide is almost non-polar, so that the above single structure, to which we must attribute a polarity of the same order as that of the nitroso-group, $\mu = 1 \cdot 9$, does not suffice. We must hybridize it with $\underset{8}{N^-}{=}\underset{7}{O^+}$ of the opposite polarity, whence the increased stability apparently offsets any tendency to dimerization. We may notice that the hybridization has brought about a sharing of *five* electrons in place of the original four in the double bond, and the bond strength is thus increased roughly in the proportion 5 : 4.

Oxygen forms a number of anions: O^- present in many metallic oxides (p. 42); O_2^- in peroxides; and O_2^-, with 17 electrons, in superoxides, such as KO_2 (p. 50). Unlike the first two ions O_2^- is coloured deep yellow and paramagnetic. The structure of O_2^- is clearly $\underset{8}{O^-}{-}\underset{8}{O^-}$; the O, O length ($1 \cdot 49$)* is slightly greater than that in H_2O_2 ($1 \cdot 47$) probably owing to the repulsion of the charges. Superoxide O_2^- must be considered as a hybrid in which the congruent forms $\underset{7}{O}{-}\underset{8}{O^-}$ and $\underset{8}{O^-}{-}\underset{7}{O}$ take an equal part, and in which the bond finally contains 3 electrons in place of two in each of the contributing structures (cf. NO above). The resulting increase of strength is reflected in the contracted O, O length ($1 \cdot 28$). A summary of O, O lengths is of interest:

$$O_2, \ 1 \cdot 207 \qquad O_3, \ 1 \cdot 28 \qquad (\text{p. 134}). \qquad H_2O_2, \ 1 \cdot 47$$
$$O_2^-, \ 1 \cdot 49 \qquad O_2^-, \ 1 \cdot 28\dagger$$

The structures CO_2, N_2O, $-CNO$, $-NCO$, $-N_3$

(1) *Carbon dioxide.*

The evidence of its infra-red spectrum is that this molecule is linear and symmetrical (OCO). The classical assumption of the

---

* Abrahams and Kalnajs, *Acta Cryst.* **7**, 838 (1954).

† *Idem. ibid.* **8**, 503 (1955).

tetrahedral distribution of the carbon valencies leads also to the formula O=C=O. From the spectrum we also learn that the O, O distance is 2·32 A. The molecule is completely non-polar.

(a) From the electronic distribution O C O we can at once see
$$6 \quad 4 \quad 6$$
that the classical bond-diagram above gives completed octets to all the three atoms. The distribution fully written out shows full symmetry $:: O :: C :: O ::$

(b) An alternative bond-diagram of less symmetry may be constructed by combining CO with O:

first, $O^+{\equiv}C^- \quad O \quad \rightarrow \quad O^+{\equiv}C \quad O^- \quad \rightarrow \quad O^+{\equiv}C{-}O^-$
$$\qquad\quad 8 \qquad 6 \qquad\qquad\quad 7 \quad 7$$
(as in F—F)

and secondly, $O \quad C^-{\equiv}O^+ \quad \rightarrow \quad O^-{-}C{\equiv}O^+$
$$\qquad\qquad 6 \quad 8$$

Having now set out all the possible bond schemes we hybridize them. The two in the set (b) are congruent, and hence will appear with equal coefficients a (p. 119). This means that the very strong polarity of the individual forms (b) will *exactly* cancel, and, as in formula (a) the separate polarities of C=O also cancel, the final hybrid is necessarily non-polar. The influence of the (b) constituents is clearly shown by the bond-lengths. The length of C=O is 1·22 A. (Table 5), and the total length of CO_2 should therefore be 2·44 A. if only formula (a) were present: it is actually 2·32 A. From the heats of formation of propylene, $CH_3.CH{=}CH_2$ (+ 5·4 Cal.), acetaldehyde, $CH_3.CH{=}O$ (− 45·8 Cal.) and allene, $CH_2{=}C{=}CH_2$ (+ 46·5 Cal.), we may infer that the heat of formation of O=C=O should be approximately 46·5 − 2 (45·8 + 5·4) = − 55·9 Cal. The heat of formation of carbon dioxide is actually − 94·0 Cal., and if we may assume that, with the exception of CO_2, all the above compounds are properly represented by a *single* structure, the 'resonance' energy of CO_2 amounts to about 38 Cal.

(2) *Nitrous oxide.*

The study of its infra-red spectrum has now placed beyond doubt the unsymmetrical formula NNO (p. 76). Like CO_2 the molecule is non-polar.

(a) On the analogy of CO_2 the grouping $\underset{6}{N^-}\ \underset{4}{N^+}\ \underset{6}{O}$ leads at once to the diagram $N^-\!\!=\!\!N^+\!\!=\!\!O$.

(b) Combination of N_2 with O yields, by analogy of CO uniting with O, $$N\!\equiv\!N^+\!\!-\!\!O^-.$$

(c) Combination of NO with N: the 'lone' electron on the nitrogen of NO being first transferred to the second nitrogen, we form the configuration $\underset{6}{O}\!\!=\!\!\underset{6}{N^+}\ \underset{6}{N^-}$, similar to $\underset{6}{O}\ \underset{6}{O}$. The union finally gives $O\!\!=\!\!N^+\!\!=\!\!N^-$, identical with (a). On hybridization all the individual polarities cancel. The case of nitrous oxide is particularly instructive in relation to the principle of resonance since, all the forms being strongly polar, only a hybridization can achieve the necessary non-polarity of the finished molecule.

(3) *The isocyanate group—NCO.*

The grouping $\underset{1+5}{-\!\!N}\ \underset{4}{C}\ \underset{6}{O}$ is so clearly similar to that in carbon dioxide that we may adopt at once the diagram $-\!\!N\!\!=\!\!C\!\!=\!\!O$.

(4) *The cyanate group—CNO.*

Here we have $\underset{1+4}{-\!\!C}\ \underset{5}{N}\ \underset{6}{O}$, and after converting to $\underset{1+5}{-\!\!C^-}\ \underset{4}{N^+}\ \underset{6}{O}$ we write $-\!\!C^-\!\!=\!\!N^+\!\!=\!\!O$. In —NCO and —CNO, as in CO_2, the effect of further forms of electronic distribution will be to shorten the bonds. Actual values in CH_3NCO are $N, C = 1\cdot19$, $C, O = 1\cdot18$.[*]

(5) *The azide group and ion.*

X-ray analysis of inorganic and organic azides shows that both the group and ion are linear (III, p. 80).

$\underset{1+5}{-\!\!N}\ \underset{5}{N}\ \underset{5}{N}$ is transformed to $\underset{1+5}{-\!\!N}\ \underset{4}{N^+}\ \underset{6}{N^-}$, and then from analogy with the two forms of CO_2 we write

(a) $-\!\!N\!\!=\!\!N^+\!\!=\!\!N^-$,   (b) $-\!\!N^-\!\!-\!\!N^+\!\!\equiv\!\!N$,   (c) $-\!\!N^+\!\!\equiv\!\!N^+\!\!-\!\!N^-$.

The last form (c) would be very unstable and readily pass into (a) (cf. p. 135). The final form of the azide radical is therefore a hybrid mainly of (a) and (b), which like CO_2 is non-polar.

(6) The constitution of diazo-methane may be arrived at by

[*] Brockway, Eyster and Gillette, *J. Amer. Chem. Soc.* **62**, 3236 (1940).

exchanging O in N_2O for the isoelectronic group CH_2. The two possible structures will be $CH_2{=}N^+{=}N^-$ and $CH_2^-{-}N^+{\equiv}N$. The C, N distance is 1·34 and N, N is 1·13, in agreement with this formulation, but affording no support to the older formula

$$CH_2{\Big\langle}{\overset{N}{\underset{N}{\|}}}\cdot$$

We may here emphasize that the degree of stabilization of molecules achieved by quantum-mechanical resonance depends upon the equality or near-equality of the energies of formation of the participating structures. From the data on p. 129 and in Table 11, we can compute approximately the energies of formation (from atoms) of the three structures contributing to the molecule of carbon dioxide: (a) $O{=}C{=}O$, 343 Cal., (b) $O^-{-}C{\equiv}O^+$ and (c) $O^+{\equiv}C{-}O^-$, each $256 + 82 = 338$ Cal. The exact equality of the energies of (b) and (c) and their near-equality to that of (a) ensures the large resonance energy indicated on p. 129. For the nitrite ion, NO_2^-, the nitro-group, $-NO_2$ and ozone (p. 134) the participating structures are congruent, like (b) and (c) above, and hence have exactly equal energies of formation, leading to a maximum resonance stabilization.

The sequence in magnitude of the following force constants (units 10^5 dynes/cm.) for the C, O group corroborates the arguments on the constitution of CO and CO_2 already given:

$H_2C{:}O$, 12·3; $CH_2{:}C{:}O$, 12·3; CO_2, 15·5; CO, 18·9.

The 8-electron group

(1) *Macromolecular structures* (cf. Chapter III, p. 40).

Compounds of the analytical composition BeO, or SiC, as well as diamond, are found to have the zinc-blende or the wurtzite structure, which can only arise when an aggregate of eight valency electrons is available, although these need not be contributed equally by the atoms (p. 53). The layer structure of boron nitride (composition BN) closely resembles that of graphite (fig. 54, p. 200). The hexagonal rings are composed of alternate atoms of B and N, with B, N = 1·446, a distance notably shorter than that expected for the single bond B—N (1·58 A.). The diatomic molecule C_2 exists in carbon vapour at high temperature, and gives rise to the Swan band-spectrum; its bond-diagram is not certain, but the C, C distance (1·312 A.) is appropriate to C=C (see Table 10, p. 64 and p. 190).

(2) *Ammonia, water and hydrogen fluoride.*

The pyramidal form of NH_3 and the triangular shape of H_2O have already been mentioned (Chapter III). It will be recalled that the p-orbits which the bonding electrons occupy in the nitrogen and oxygen atoms are mutually directed at right angles (p. 108). We might therefore expect that the bonds linking hydrogen with these atoms would be similarly disposed. However, the protons exert a strong mutual repulsion, and the angle between the bonds may be widened to approximately the 'tetrahedral' angle (about 109°). A calculation on p. 46, Chapter III, indicates that HCl is not to be regarded in the anhydrous state as an ion-pair; similar considerations apply to HF.

(3) *The ions* $(NH_4)^+$ *and* $(H_3O)^+$.

The ammonium ion arises from the reaction

$$NH_3 + H^+ \rightleftharpoons (NH_4)^+.$$

To follow the formation electronically we transfer one electron from the nitrogen atom to the proton, giving the configuration

$$(H_3) \quad \underset{7}{N^+} \quad \underset{1}{H} \quad (cf. \underset{7}{F} \quad \underset{1}{H})$$

and then on forming the bond we have the ion $(NH_4)^+$, in which the formal positive charge is upon the central nitrogen, and all the N—H bonds are equivalent. The four bonds are tetrahedrally arranged, in consonance with the optical activity of suitably substituted ammonium salts (III, p. 33). The ion H_3O^+ is commonly called the 'hydrogen' ion. It arises from the reaction $H_2O + H^+ = H_3O^+$. The transference of one electron from oxygen to the proton gives $\underset{7}{H_2O^+} \underset{1}{H}$ and the completed ion contains positively charged oxygen and three equivalent hydrogen atoms. The constitution of XH_4 and XH_3 structures may be deduced more directly. It may be presumed that B^-, C and N^+ have identical electronic configurations, viz. $2s^2 2p^2$, in the ground-state, and that all these atomic structures became quadrivalent by promotion and hybridization to the configuration $[2s\,2p^3]$. In this state they yield respectively $(BH_4)^-$ (see p. 61), CH_4, and $(NH_4)^+$, as well as $(BF_4)^-$ and CF_4. In all these compounds the directions of the four bonds are tetrahedral, as in C compounds. Further, N and O^+ both have the configuration $2s^2 2p^3$, and thus $(OH_3)^+$ is analogous to NH_3 as regards electronic constitution, and the *pyramidal* relationship of the three bonds.

The hydrolytic activity of H_3O^+ is mainly traceable to intimate contact between the positive ion and the point of attack in the hydrolysable molecule, as in the following scheme for the hydrolysis of the isonitriles:*

$$
\begin{array}{l}
\text{CH}_3 \\
\mid \\
\text{N}+ \\
\text{H} \quad \parallel \quad \text{H} \quad \text{H} \qquad \text{N—H} \\
\qquad\qquad\qquad\qquad\qquad \longrightarrow \quad \text{H}+\text{HO—C}=\text{O}+2\text{H}^+ \\
+\text{O}\ldots\ldots\text{C}\ldots\ldots\text{O}+ \\
\text{H} \quad \text{H} \quad \text{H} \qquad\qquad\qquad \text{H}
\end{array}
$$

It must not be concluded that in the ion H_3O^+, and in other examples in which the formation of a co-ionic bond seems to place ionic charges upon atoms so bonded, that the *actual* charge attains the full possible value (4·80 e.s.u.). Attention has already been drawn (p. 124) to the case of carbon monoxide $C^-\!\!\equiv\!O^+$, where the 'natural' polarity $C_+\!\!\equiv\!O_-$ almost completely offsets the *formal* charges arising from the co-ionic bond. It is by no means easy to calculate the actual charges in specific cases, but it is clear that in H_3O^+, for example, the polarity $O_-\!\!-\!H_+$ will considerably reduce the formal positive charge on the oxygen atom.

The amine-oxides and the corresponding bases $(R_3N.OH)^+.OH^-$ are closely connected with the $(NH)_4^+$ ion. The interaction of trimethylamine with aqueous hydrogen peroxide yields the ionic hydroxide, $(Me_3NOH)^+.OH^-$, which on gentle dehydration

---

* Cf. Lowry, *Chem. Ind.* **42**, 43 (1923); Bronsted, *Rec. trav. chim.* **42**, 718 (1923); *J. Physical Chem.* **30**, 777 (1926); *Trans. Faraday Soc.* **24**, 630 (1928).

gives the oxide $(CH_3)_3NO$. The electronic structure of the latter can only be obtained by uniting the grouping $(CH_3)_3\overset{7}{N^+}\ \overset{7}{O^-}$ to give $(CH_3)_3N^+$—O^-. The high m.p. of the oxide (208°) strongly supports its polar constitution. The length of the N, O bond is $1\cdot36\,A$.* and its force constant is approximately 5×10^5 dynes/ cm.† (pp. 39, 64).

(4) $BF_3.NH_3$.

This compound was discovered by Thénard in 1809, and its properties examined by J. Davy in 1812. It is produced when the gases NH_3 and BF_3 are brought into contact, and the resulting solid is treated with HCl. It is colourless, and sublimes with dissociation (v.d. 23); its stability to acid is shown by its method of preparation. Its electronic formula is arrived at as follows:

$$(F_3)\ \overset{7}{B^-}\ \ \overset{7}{N^+}\ (H_3) \quad F_3B^-\!\!\!-(NH_3)^+.$$

It is interesting to note that although in BF_3 and BCl_3 the boron atom only attains a sextet of electrons, in the above compound, and in the stable ion BF_4^-, its octet is completed. The structures of $BH_3.CO$ and of $BH_3.Me_3N$ are doubtless similar (Chapter III, p. 60).

The 18-electron group

The nitrite ion NO_2^- and the nitro-group —NO_2 have already been discussed.

As further examples we take *nitrosyl fluoride* and *ozone*.

The fluoride is F—N=O, formed by combining the configurations $\overset{7}{F}\ \ \overset{7}{N}=O$.

From $\overset{6}{O}\ \ \overset{8}{O}=\overset{8}{O}$ we form $\overset{7}{O^-}\ \ \overset{7}{O^+}=O$, and find for ozone

$$O^-\!\!\!-O^+\!=\!O,$$

a triangular molecule like NO_2; and as in the latter, the O, O bonds are equalized by resonance between congruent forms.

---

* Lister and Sutton, *Trans. Faraday Soc.* **35**, 495 (1939).

† Goubeau and Fromme, *Z. anorg. Chem.* **258**, 18 (1949).

The 24-electron group

The ions BO_3, CO_3 and NO_3.

X-ray analysis of borates, carbonates and nitrates shows that all these ions of the type XO_3 have the same shape—an equilateral triangle, with X at its geometrical centre; all X, O distances are equal in a given ion. The ion CO_3 is formed by the union of CO_2 with O^- as in the actual reaction

$$CO_2 + Ca^{++}.O^- = Ca^{++}.CO_3^-.$$

CO_2 in the electronic form $O{=}C{=}O$ cannot, however, react in this way, as all the electrons of the carbon atom are already engaged in bonding. We therefore transfer one electron-pair in one bond to oxygen, producing the (activated) structure $O{=}C^+{—}O^-$, which can now react with O^- in the normal way of forming a co-ionic bond:

$$O{=}\underset{8}{C}{—}O^-$$

8   6   8

$$O{=}\overset{|}{C}{—}O^-$$
$$\underset{O_-}{|}$$

This structure as it stands, although planar in accordance with classical tetrahedral theory, is obviously in disagreement with the experimental fact that all C, O distances are equal. We can construct, however, two other (congruent) forms:

$$O^-{—}\underset{\overset{\|}{O}}{C}{—}O^- \quad \text{and} \quad O^-{—}\underset{\overset{|}{O_-}}{C}{=}O$$

and in the resonance hybrid all partake equally. The ion SO_3^- (and ClO_3^-) similarly forms a resonating group of three members, leading to equalization of the X, O distances (p. 85). In the formula for CO_3^- C may clearly be replaced by the isoelectronic units N^+ and B^-, giving respectively

$$O^-{—}\underset{\overset{|}{O_-}}{B^-}{=}O \quad \text{and} \quad O^-{—}\underset{\overset{|}{O_-}}{N^+}{=}O$$

The structure above for BO_3^- raises the question of the stability of the bond $A^-{—}B^-$, or $A^+{—}B^+$ as in one form of the azide ion (p. 130). Pauling* discusses this important matter along the following lines. The work $+W_i$ of forming the ions A^+ and B^- is

* *J. Amer. Chem. Soc.* **59**, 17 (1937).

(ionization energy of A) − (electron affinity of B); the energy − W_b released in respect of the charges on forming the bond is their mutual coulombic energy. Hence if $W_i \sim W_b$, the co-ionic bond A^+—B^- has about the same stability as A—B. It is therefore clear that in the above cases of *like* adjacent charges in which the coulombic energy W_b is positive, both types of bonding are much less stable than the ordinary covalent bond. Pauling proposes a rule of 'adjacent charge' precluding the formation of these types of bond.* Applying this rule to the structure for BO_3^- above, we shall expect it to pass into

$$\begin{array}{ccc} O^- & & O^- \\ & \diagdown \!\!\! B \!\!\! \diagup & \\ & | & \\ & O_- & \end{array}$$

in which the boron atom has only a sextet of electrons, and the ion would be planar like BF_3.† The structure for the nitrate ion is also the resultant of a resonating group like that of CO_3^-. The same structure could be reached by combining the nitrite ion (p. 127) with O. All the electrons on the central atom being concerned in bonding, the ions BO_3, CO_3 and NO_3 represent a final stage in oxidation (cf. SO_3^-, ClO_3^- and BrO_3^-).

Complete resonance between the three structures of the CO_3 and NO_3 groupings is only possible as long as they remain independent units, i.e. ions. In the unionized alkyl nitrates resonance is confined to only two oxygen atoms, as in the nitro-group (see above, p. 128):

$$CH_3\!\!-\!\!O\!\!-\!\!N^+\!\!\!\diagup^{\displaystyle O}_{\displaystyle O_-} \quad \text{and} \quad CH_3\!\!-\!\!O\!\!-\!\!N^+\!\!\!\diagup^{\displaystyle O^-}_{\displaystyle O}$$

and it is found that the O—N bond towards the CH_3 is longer than the other equal N, O distances.‡ Hence methyl nitrate may be regarded as O-nitro-carbinol (cf. HIO_3, p. 97).

* Notable exceptions occur in N_2O_4, p. 127 and in the ion O_2^-, p. 128.
† The length B—F is well established—as 1·309 in the 'spectroscopic' molecule BF, and as 1·29–1·30 by electron-diffraction and spectroscopic methods applied to BF_3. Taking O— as 0·08 greater than F— we calculate for B—O 1·38. The value observed in borates is 1·36. On the other hand, C, O in CO_3^- (1·31) is intermediate between C—O, 1·42 and C=O, 1·20.
‡ Pauling and Brockway, *J. Amer. Chem. Soc.* 59, 13 (1937).

The influence of resonance upon the stability of chemical structures is well exemplified by comparing the energy of reduction of the nitrate ion with that of methyl nitrate. For the reduction

$$\left[O{=}N^+{<}{}^{O^-}_{O^-} \right]^- \rightarrow \left[O{=}N{\diagdown}_{O^-} \right]^- + O$$

89·5 Cal. are required: for the corresponding change

$$CH_3{-}O{-}N^+{<}^O_{O^-} \rightarrow CH_3{-}O{-}N{=}O + O$$

only **75·0** Cal. are necessary. In the nitrate ion resonance can occur freely over all the N, O bonds, but is restricted to two N, O bonds in the alkyl nitrate.

Definitions of valency

It will be appropriate to conclude this preliminary account of the application of modern theories by reviewing the definitions of valency that have arisen from these theories. An essentially new situation is created by the introduction of the conception of the co-ionic bond, which appears to destroy the basic classical principle of mutual action, stated in Chapter II, p. 19, as a 'rule of valency'. The electronic states of carbon in CH_4 and of oxygen in H_2O are identical with their states in CO as regards occupancy of orbits, and any rational definition of the valencies of these elements must allow them to be the same in all three compounds. The existence of such stable compounds as the ions $BeF_4^=$, BF_4^-, and $BF_3.NH_3$ in part disallows our original electronic definition (p. 110) that the valency of an element (if necessary promoted to a valency state) is the number of 'unrestricted' electrons in the atom, and instead lays emphasis on the number of *vacant orbits* in the atom. For light atoms in Groups IV to VII inclusive, however, the number of unrestricted electrons happens to equal that of the unfilled orbits. In the past the absence of a theory of atomic structure put numerical valency into the position of a guide to the construction of bond-diagrams. Now that we can deduce these more directly a rigid definition of valency, even if it could be successfully formulated, would be to a great extent a

work of supererogation. An exponent of modern theories remarks:
'The more precisely the interactions between atoms are known,
the less need is there to introduce the idea of valency. If it were
possible to work out in practice, instead of only in principle, the
interactions and the energy of any system of atoms, the need
for introducing the concept of valency would never arise.'*

Stereochemistry of compounds of the light elements

Table 21 shows a summary of the best-established interbond
angles for simple compounds of the lighter elements. The
nucleus of hydrogen in the bond X—H is the more poorly
screened (by the bonding pair of electrons) the more polar the
linkage. For X = C, N and O, respectively, we find the following
electric moments (Table 4, p. 38):

$$C—H \quad (0\cdot4),† \qquad N_-—H_+ \quad 1\cdot31, \qquad O_-—H_+ \quad 1\cdot53.$$

Thus the nuclei of two hydrogen atoms bonded to one of the
above atoms will repel each other with a force increasing in the
order C, N, O. The nuclei of bonded halogen atoms are in all
circumstances well screened by the electron shell, and the
polarity in X_+—(hal.)_ cannot much increase this screening. However, repulsion between contiguous parts of the electron shells will
cause two bonded halogen atoms also to repel each other. In
carbon compounds such as CH_2F_2, CH_2Cl_2 and $CHCl_3$ we shall
have not only repulsion between the two pairs of like atoms but
attraction between the oppositely charged hydrogen and halogen.
Such compounds should therefore prove suitable for testing the
extent to which these interatomic forces can affect the interbond angles. If the molecular configuration were *mainly* determined by these forces we should expect a square distribution
in which like atoms are at a maximum distance apart and unlike

---

* Penney, *The Quantum Theory of Valency* (1935), p. 42.

† The polarity of the C—H bond is always small, and probably the
magnitude and even its direction (C_+—H_ or C_—H_+) depend on the
compound containing the bond; in the paraffins the bond is C_+—H_.
See p. 204 and Gent, *Quart. Rev.* 2, 383 (1948).

Table 21

(Methods: (S) Infra-red, Raman and microwave spectroscopy. (ED) Electron diffraction.)

Compounds of carbon

Difluoromethane	CH_2F_2	$\angle FCF$	$108°\ 17' \pm 6'$	(S) Lide, *J. Amer. Chem. Soc.* **74**, 3548 (1952)
Methylene chloride	CH_2Cl_2	$\angle ClCCl$	$111°\ 47' \pm 1'$	(S) Myers and Gwinn, *J. Chem. Physics*, **20**, 1420 (1952)
Chloroform	$CHCl_3$	$\angle ClCCl$	$110°\ 24'$	(S) Ghosh, Trambarulo and Gordy, *J. Chem. Physics*, **20**, 605 (1952)
Ethylene	$CH_2{:}CH_2$	$\angle HCH$	$119°\ 55' \pm 30'$	(S) Gallaway and Barker, *J. Chem. Physics*, **10**, 88 (1942)
Formaldehyde	$H_2\overset{..}{C}{:}O$	$\angle HCH$	$120° \pm 1°$	(ED & S) Stevenson, LuValle and Schomaker, *J. Amer. Chem. Soc.* **61**, 2508 (1939); Davidson, Stoicheff and Bernstein, *J. Chem. Physics*, **22**, 289 (1954)
Ketene	$H_2C{:}C{:}O$	$\angle HCH$	$123·3° \pm 1·5°$	(S) Arendale and Fletcher, *J. Chem. Physics*, **21**, 1898 (1953)

Compounds of nitrogen and other Group V elements

Ammonia	NH_3, ND_3		$106°\ 47'$	(S) Dennison, *Rev. Mod. Physics*, **12**, 175 (1940)
Nitrogen trifluoride	NF_3		$102°\ 9'$	(S) Sheridan and Gordy, *Phys. Rev.* **79**, 513 (1950)
Trimethylamine	$(CH_3)_3N$	$\angle CNC$	$108° \pm 4°$	(ED) Brockway and Jenkins, *J. Amer. Chem. Soc.* **58**, 2036 (1936)
Phosphine	PH_3, PH_2D, PHD_2		$93°\ 18'$	(S) Sirvetz and Weston, *J. Chem. Physics*, **21**, 898 (1953)
Arsine	AsH_3, AsD_3		$89°\ 50'$	(S) McConaghie and Nielson, *Phys. Rev.* **75**, 633 (1949)
Phosphorus trifluoride	PF_3		$104° \pm 3°$	(S) Gilliam, Edwards and Gordy, *Phys. Rev.* **75**, 1014 (1949)
Phosphorus trichloride	PCl_3		$100°\ 6' \pm 20'$	(S) Kisliuk and Townes, *J. Chem. Physics*, **18**, 1109 (1950)

Compounds of oxygen and other Group VI elements

Water	H_2O, D_2O		$104°\ 27'$	(S) Dennison, *Rev. Mod. Physics*, **12**, 175 (1940)
Oxygen fluoride	F_2O		$101°\ 30' \pm 1·5°$	(S) Bernstein and Powling, *J. Chem. Physics*, **18**, 685 (1950)
Oxygen chloride	Cl_2O	$\angle COC$	$110°\ 48' \pm 1°$	(ED) Dunitz and Hedberg, *J. Amer. Chem. Soc.* **72**, 3108 (1950)
Dimethyl ether	$(CH_3)_2O$	$\angle COC$	$111° \pm 3°$	(ED) Sutton and Brockway, *J. Amer. Chem. Soc.* **57**, 473 (1935)
Dioxane	$O\langle\overset{\displaystyle CH_2CH_2}{CH_2CH_2}\rangle O$	$\angle CCO$ $\angle COC$	$109°\ 30' \pm 5°$ $112° \pm 5°$	(ED) Allen and Sutton, *Acta Cryst.* **3**, 46 (1950)
Hydrogen sulphide	H_2S		$92°\ 16'$	(S) Cross, *Phys. Rev.* **46**, 536 (1934); *ibid*, **47**, 7 (1935)
Hydrogen selenide	H_2Se		$91°$	(S) Palik, *J. Chem. Physics*, **23**, 980 (1955)
Sulphur dichloride	SCl_2		$100°\ 20'$	(S) Stammreich, Forneris and Sone, *J. Chem. Physics*, **23**, 972 (1955)

atoms closer than in the tetrahedral configuration. Table 21 above, however, indicates that the variation from the tetra-hedral angle 109° is small in the three halo-genated methanes. In compounds with 'double-bonded' carbon, the hybridized car-bon orbits are of the trigonal type mentioned in p. 115. The expected angle 120° appears in ethylene and formaldehyde, and, less exactly, in ketene. In the compounds of carbon therefore it may be inferred that it is the valency state of this atom which primarily determines the interbond angles.

In ammonia and water larger forces will exist, owing to the greater polarity. If we assume that in these hydrides the bonding orbits are pure p-orbits of nitrogen or oxygen, then the (undis-torted) interbond angle should be 90° in both NH_3 and H_2O (p. 108). The actual angles are respectively 107° and 104° (Table 21). An approximately equal angle is seen in F_2O, in which as explained above the effect of polarity upon halogen repulsion will be small. The angles in the hydrides of phosphorus, arsenic, sulphur and selenium (Table 21) depart little from the theo-retical angle of 90°, facts which might be explained by the greater separation between the protons occasioned by the larger central atom, and by the lessened polarity of the linkages:

Atomic radii	N	0·73	O	0·74
	P	1·10	S	1·04
	As	1·25	Se	1·16

Moments $P_- \text{—} H_+$ 0·36, $As_- \text{—} H_+$ 0·10, $S_- \text{—} H_+$ 0·68.*

Another explanation of the characteristic interbond angles, based primarily upon the relationships of the orbits in the central atom, has come into favour.† All the available evidence, especially that of the proved optical activity of substituted ammonium compounds (p. 33), supports the equivalence and tetrahedral direction of the four bonds in NH_4^+. It follows that

* Smyth, *J. Physical Chem.* **41**, 209 (1937).

† Coulson, *Valence*, Oxford 1952, pp. 207 *et seq.*

the s-orbit of the nitrogen atom, which comes into play in these compounds, must undergo hybridization with the p-orbits to form 'tetrahedral' orbits similar to those in carbon. It therefore is not impossible that such a process of hybridization has also occurred in ammonia, in which one of the four hybrid orbits contains paired electrons and is non-bonding. A similar electronic constitution can be assumed for 3-covalent oxygen, as in OH_3^+ and the oxonium compounds. Now a hybrid orbit, unlike its s and p constituents is not symmetrical about the atomic nucleus (see fig. 31, p. 115), and hence the centre of action of its (negative) electronic charges does not coincide with the nucleus: in short, a fully occupied hybrid orbit creates an 'atomic' dipole, independent of any dipoles supposed to exist in the chemical bonds of the molecule. These conceptions applied to bicovalent oxygen would imply that in, for example, H_2O there are two such atomic dipoles directed in a plane perpendicular to that of the OH bonds. In NH_3 and H_2O the resultant atomic dipole reinforces that of the bonds, but calculation has shown that much of the measured dipole moments could arise from the atomic dipoles alone. The small moment of NF_3 (0.2D) is neatly explained by the *opposition* of the bond and atomic dipoles.

Attention may be drawn to the following compounds, in which unusual interbond angles are found:

*cyclo*Propane	CH$_2$—CH$_2$ \\ / CH$_2$ (b.p. $-34°$)	$\angle$ CCC $\angle$ CCH $\angle$ HCH	$60°$ $116.4°$ $118.2°$
		Hassel and Viervoll, *Acta Chem. Scand.* **1**, 149 (1947)	
*cyclo*Butane	CH$_2$—CH$_2$ | | CH$_2$—CH$_2$ (b.p. $12°$)	$\angle$ CCC	$90°$
Ethylene oxide	CH$_2$—CH$_2$ \\ / O (b.p. $14°$)	$\angle$ COC $\angle$ OCC $\angle$ HCH	$61.4°$ $59.3°$ $116.7°$
		Cunningham, Boyd, Myers, Gwinn and Le Van, *J. Chem. Physics*, **19**, 676 (1951)	
Phosphorus	P$_4$		$60°$
α- and β-quartz	SiO$_2$	$\angle$ SiOSi	$142°$
p,p-Di-iodophenyl ether	(IC$_6$H$_4$)$_2$O	$\angle$ COC	$123° \pm 2°$

*cyclo*Propane differs little in chemical properties from propylene $CH_3.CH:CH_2$. In being unaffected in the cold by bromine or hydrogen iodide *cyclo*butane shows less olefinic character. With cold hydrochloric acid ethylene oxide yields glycol chlorhydrin, the change being reversed by alkalis. Ethylene oxide also reacts rapidly with bromine, and by yielding CHI_3 with hypoiodites shows its tendency to change to acetaldehyde, CH_3CHO.

If the bond energies in *cyclo*propane and ethylene oxide were normal, their energies of formation from atoms would amount to 834 and 648 Cal./mol. respectively (Table 8). The energies calculated from the heats of combustion (499·6* and 312·5 Cal.†) are in fact 812·4 and 621·4 Cal./mol. The differences (*ca.* 25 Cal.) represent the total 'strain' in all the bonds of these molecules (see further, p. 203).

* Kaarsemaker and Coops, *Rec. trav. chim.* **71**, 261 (1952).
† Crog and Hunt, *J. Physical Chem.* **46**, 1162 (1942).

Chapter V

FURTHER DEVELOPMENT OF THE ELECTRONIC THEORY AND ITS GENERAL APPLICATION

In discussing the light elements He to Ne it has been necessary to recognize only two types of atomic electronic orbit, s- and p-orbits; and to make between these types only a spatial distinction, viz. that s-orbits are spherically symmetrical and p-orbits disposed with their axes mutually at right angles (p. 107 and fig. 28). For a clear understanding of the atomic structures and resulting valencies of the heavier elements we shall need not only to recognize further types of orbit, but to distinguish them in greater detail. In this elaboration we shall continue to rely much on the results of spectroscopy.

It will be recalled (p. 105) that the emission of a spectral 'line' (visible or invisible) is the result of an electron transition (or 'jump') from one orbit to another of less energy—a 'lower' or 'deeper' orbit. The frequency ν of the emitted radiation is related to the energy difference ΔE by the fundamental quantum equation $\Delta E = h\nu$. It is one of the chief tasks of practical spectroscopy to identify and evaluate the various energy-levels, lying between the lowest or ground-state and complete ionization, within which limits transitions from one level to another occur. The final result for a particular atom or ion is commonly represented graphically in what is called a *term-scheme*. Such a representation for the He atom is found in fig. 32, and below (fig. 33) is seen the term-scheme for hydrogen. In view of the simplicity of the hydrogen atom it is not surprising that the scheme and the spectrum related to it are both also very simple.

With the ground-state as zero, the energy-levels in hydrogen as we ascend the array of terms towards the highest or ionization level are given by the expression $E_n = \dfrac{2\pi^2 e^4 m}{h^2 n^2} = \dfrac{Rh}{n^2}$; in this, n, termed the *principal quantum number*, takes the successive integral values 1, 2, 3, etc.; R, which is the same for all levels, is called Rydberg's constant. The spectral frequencies of hydrogen. will hence be

$$\frac{\Delta E}{h} = \nu = R\left(\frac{1}{n_i^2} - \frac{1}{n_n^2}\right),$$

where n_i and n_n are two integers. It had been recognized early in the development of spectroscopy that spectral frequencies could be expressed as the difference of two *terms* in this way, and after the quantum theory gave to the 'term' the physical significance of energy-level, the name was retained with the new meaning.

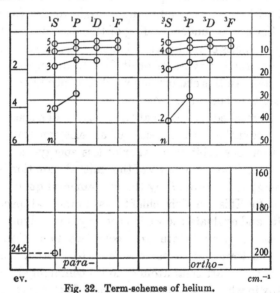

Fig. 32. Term-schemes of helium.

[Energies of levels in electron-volts (ev.) and in wave-numbers × 10⁻³.]

If for the moment we picture in hydrogen the electron as a point-charge moving in the field of the central proton, we at once notice that an infinite number of orbits could be conceived, of varying eccentricity (including circular), to comply with the sole condition that the energy of the electron-proton system remains constant. Each orbit would, however, be distinguished by its own special relation of kinetic energy (associated with angular momentum) to potential energy. Only in the *circular* orbit do we find the energy constantly divided equally between these two forms; in *elliptical* orbits the ratio between the forms of energy changes with the position of the electron in the orbit. It is, however, repugnant to the idea of fixed atomic properties that a planetary electron should have, so to speak, an unfettered choice of orbit. Bohr, basing his ideas upon the quantum theory of Planck, was the first to suggest the severe limitations: (1) that angular momentum (moment of inertia × angular velocity = $I\omega$) must always be an integral multiple of the universal quantity $h/2\pi$: thus $I\omega = l \cdot h/2\pi$, where $l = 1, 2, 3$, etc.; (2) that the maximum permissible value of the *second* or *orbital* quantum number l does not exceed that of n, the *principal quantum number*. Developments of modern theory have in general confirmed and conferred physical significance upon these, at first arbitrary assumptions, and finally shown that the angular momentum is given by $\sqrt{l(l+1)} \cdot h/2\pi$ in place of $l \cdot h/2\pi$, and that the maximum permissible value of l is $(n-1)$.

Table 22. *Classification of orbits*

n, principal quantum number	1	2		3			4				etc.
l, orbital quantum number	0	0	1	0	1	2	0	1	2	3	etc.
Types of orbit available	s	s	p	s	p	d	s	p	d	f	

Orbits we have previously labelled s-orbits are therefore those in which the electron has no angular momentum; p-orbits ($l = 1$) have one unit. The existence of 'orbits' without angular momentum is of course incompatible with Bohr's semi-classical theory, in which classical centrifugal force was invoked to prevent the planetary electron from falling into the nucleus. Like 's' and 'p' the labels 'd' and 'f' are drawn from the nomenclature of classical spectroscopy (see p. 148). Owing to the fact that in the hydrogen atom the electron is always submitted to a force varying according to the inverse square law the energy of the electron in hydrogen orbits depends only upon the principal quantum number n. This extreme simplicity is lost even when only two electrons are present, as in He (fig. 32), and in the term-diagrams (fig. 33) for the 'hydrogen-like' atoms, Li, Na and K, it is evident that although the energies of orbits can still be grouped together under the number n, there is a continuous difference (increase) of energy through the series s, p, d, f. Modern theory solves the question of how many equivalent orbits of a given type may be distributed spatially. The number m is found to be $m = 2l + 1$. As we have noticed qualitatively before there is therefore only one s-orbit in any shell (the spherical symmetry would preclude

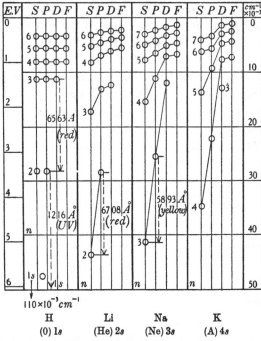

Fig. 33. Term-schemes of hydrogen and the hydrogen-like atoms Li, Na and K.

another); and there are $2 + 1 = 3$ p-orbits, arranged with their axes mutually at right angles. There will be five d-orbits, and seven f-orbits at levels where these orbits are permissible (i.e. n not less than 3). The spatial disposition of d- and f-orbits is complex and will not concern us.

The applications of the theory of quantum mechanics have in recent years explained and confirmed well-known and fortunate limitations on the complexity of atomic spectra. No spectral frequency is ever found that is related (by the quantum equation $\Delta E = h\nu$, p. 144) to an electronic transition between orbits of the *same l* type. Transitions occur between *s*- and *p*-orbits, *d*- and *f*-orbits, etc., but never between two *s*-orbits, or two *p*-orbits, etc. The limitation may be summarized in the 'selection rule' $\Delta l \pm 1$.*

The assignment of prominent spectral frequencies to spectral series is conventionally effected under the following classification:

$np \rightarrow (1)s$	Principal series	$nd \rightarrow (2)p$	Diffuse series
$ns \rightarrow (2)p$	Sharp series	$nf \rightarrow (3)d$	Fundamental series

The origin of the first 'line' of the principal series for Li (6708 A.) and for Na (5893 A.) is indicated in fig. 33. It is from the initial letters of the description of the series that the nomenclature of orbits is taken.

The quantization of angular momentum is also applied to electron-spin (p. 109). We have already seen that the conception of electron-spin is required to explain the fact, ascertained from spectroscopy, that every orbit can be doubly occupied. In the quantization of angular *orbital* momentum we introduce the unit $h/2\pi$, and it is reasonable to suppose that an equal unit will be suitable for spin momentum. We must however so choose our spin quantum number s that it can take only two equal values, of opposite sign (corresponding to the two directions of spin) and differing of course by unity. A little consideration shows that the only quantum numbers possible are $\frac{1}{2}$ and $-\frac{1}{2}$. It follows that the angular momentum of the spinning electron is constant in magnitude. We may now re-state Pauli's principle (p. 108) more precisely: *if two or more electrons in an atom possess identical quantum numbers n, l and s, each electron must occupy a different one of the 2l + 1 orbits available; one such orbit can hold at most 2 electrons, with s = $\frac{1}{2}$ and $-\frac{1}{2}$.*

It is convenient to be able to summarize compactly the electronic state, and particularly the ground-state, of an atom containing more than one electron. Electrons in atoms such as H, He, Li and Be, which in their ground-state contain only electrons in *s*-orbits, obviously cannot possess any resultant orbital angular momentum L. These atoms are said to have S ground-states. Of course on excitation other states will arise.

---

* This rule ceases to be rigorous for transitions involving changes in the states of more than *one* electron.

An atom which contains p- as well as s-electrons may also assume an S ground-state. Thus in Ne $(1s^2 2s^2 2p^6)$ the symmetry of the orbits ensures a complete absence of resultant angular momentum. Any atom containing 2, 3 or 4 p-electrons (e.g. carbon, nitrogen or oxygen) may also by symmetry pass into an S state, although, owing to the interaction of orbital momentum with spin momentum mentioned below, the S state may often not be the actual state of lowest energy, i.e. the ground-state. Atoms containing an incomplete group of p-, d- or f-electrons commonly adopt a ground-state with one or more units $(h/2\pi)$ of resultant orbital momentum—one unit $(L = 1)$ gives a P state, two $(L = 2)$ a D state, and three $(L = 3)$ an F state. If an element contain in its outer shell a completed group (e.g. ns^2, np^6) and one other electron, p-, d- or f-, then the state must obviously be solely determined by this electron, and be P, D or F respectively (cf. scandium, below). Also, if an atom contain an outer p-, d- or f-group *deficient* in only one electron (e.g. np^5, as in F and Cl, or nd^9) then the ground-state is necessarily P, D or F respectively.

The spin momentum may also by symmetry be partially or wholly self-neutralizing. If the *resultant* spin momentum (in units of $h/2\pi$) is S (not to be confused with the *state* symbol), this number may be usefully attached at the left of the state symbol, but for special reasons spectroscopists have found it more convenient to place in this position the number $2S + 1$, termed the *multiplicity*. Since an electron possesses a constant angular momentum of $\frac{1}{2}$ unit, the number $2S$ registers the number of electrons with unrestricted spin (see p. 110), and hence the minimum chemical valency. Lastly, the resultant orbital momentum (L) and spin momentum (S) may also couple to give a composite resultant (J). It has thus become customary to express in the final term-symbol of an electronic state

(*a*) the resultant *orbital angular momentum*: S, P, D, etc.;

(*b*) the *multiplicity* $2S + 1$, placed to the upper left of the symbol: ^1S, ^3P, ^2P, etc.;

(*c*) the total resultant number of units of angular and orbital momentum (J) may be attached to the bottom right of the symbol: $^3P_2$, $^4S_{\frac{3}{2}}$, $^2S_{\frac{1}{2}}$.

Table 23. *Ground-states in Series II to IV*

SERIES II	Configuration	Obs. ground-state, with possible other states for $n=2$	SERIES III	Obs. ground-state	SERIES IV	Obs. ground-state	SERIES IV (continued)	Obs. ground-state
Li	$1s^2\,2s$	^2S	Na	^2S	K	^2S	Cu	^2S
Be	$1s^2\,2s^2$	^1S	Mg	^1S	Ca	^1S	Zn	^1S
B	$1s^2\,2s^2\,2p$	^2P	Al	^2P	Sc	^2D	Ga	^2P
C	. . $2p^2$	$^3P,\,(^1D,\,^1S)$	Si	^3P	Ti	^3F	Ge	^3P
N	. . $2p^3$	$^4S,\,(^2D,\,^2P)$	P	^4S	V	^4F	As	^4S
O	. . $2p^4$	$^3P,\,(^1D,\,^1S)$	S	^3P	Cr	^7S	Se	^3P
F	. . $2p^5$	^2P	Cl	^2P	Mn	^6S	Br	^2P
Ne	. . $2p^6$	^1S	Ar	^1S	Fe	^5D	Kr	^1S
					Co	^4F		
					Ni	^3F		

(When several states are shown they are in order of increasing energy.)

Those unfamiliar with spectroscopic conventions will probably be assisted by examples of their application. We may take first the actual phenomena which historically led to the discovery of electron-spin (Goudsmit and Uhlenbeck, 1925). On excitation by the intake of radiant energy the (3)s electron in Na reaches the (3)p level above. In a p-orbit $l=1$, and, the electron possessing a spin of $s=|\tfrac12|$, the resultant total angular momentum may be either $J=1+\tfrac12$, or $J=1-\tfrac12$, according to the direction of the spin. Since the energy is only slightly changed by a change of J, the p-level is 'split' into two sub-levels close together. Such a group of nearly equal levels is termed a *multiplet*, and the *multiplicity* $(2S+1)=2$ records the number of such sub-levels. The full symbols for the *doublet* state are $^2P_{\frac32}$ and $^2P_{\frac12}$, the latter being the lower. When, as must always happen, a very large number of Na atoms are simultaneously excited, in some the electron reaches one p-level and in others the second p-level. Now in the s-level $(l=0)$ the electron-spin alone contributes to the angular momentum, and although the direction of spin is given by the sign of s, *the energy does not depend on this sign*; in short the s-level is a single state (fig. 34). It will be seen that when the electrons return in the Na atoms to their lowest s-level, *two* spectral lines of nearly equal frequency are emitted—the well-known D-lines of sodium.

Quartet states (e.g. ^4S, ^4P etc.) arise from three unrestricted electrons, each with $s = \pm \frac{1}{2}$:

Electron   1   2   3

$$s \quad \left\{ \begin{matrix} \frac{1}{2} & \frac{1}{2} & \frac{1}{2} \\ -\frac{1}{2} & -\frac{1}{2} & -\frac{1}{2} \end{matrix} \right\} \quad S = \tfrac{3}{2}, \tfrac{1}{2}, -\tfrac{1}{2}, -\tfrac{3}{2}.$$

The resultant spins S are obtained by taking the separate spins in *triads*. If we are dealing with a ^4P state ($L = 1$), the values of $|J|$, upon which the energy depends, are

$$|J| = \tfrac{5}{2}, \tfrac{3}{2}, \tfrac{1}{2},$$

giving only *three* levels of distinct energy. For a ^4D state ($L = 2$)

$$J = \tfrac{7}{2}, \tfrac{5}{2}, \tfrac{3}{2}, \tfrac{1}{2},$$

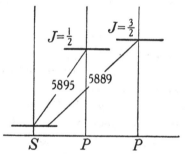

Fig. 34. Origin of the D-lines in the spectrum of sodium.
(Wave-lengths in A. units.)

and here the quartet state manifests the full number of actual sub-levels.

Effective multiplicity

True quartet state $\Big\langle$ $L = 0$————Single
$L = 1$————Triple
$L = 2$————Quadruple

Since for an S state $L = 0$, all S states, whatever the multiplicity, give *single* levels. The ground-state of N is a single level, with symbol $^4S_{\frac{3}{2}}$. Such suppression of potential multiplicity is an example of the phenomenon termed *degeneracy*.*

* The condition of degeneracy can be brought to light by causing the atom to radiate in a magnetic field, when the 'magnetic splitting' or Zeeman effect reveals the true multiplicity of the degenerate state.

Finally, let us take the ^3P ground-state of C or O, which arises from two p-electrons. For $s = \pm \frac{1}{2}$, combining in *pairs*, $S = 1, 0$ or -1. Since $L = 1$, $J = 2, 1$ and 0, and the (triplet) state shows no degeneracy. Its symbols are $^3P_2$, $^3P_1$ and $^3P_0$, the *last* being the lowest for carbon, and the *first* for oxygen ('inverted' triplet).

In Table 23 we may notice several features of chemical importance.

(1) Rather unexpectedly, in cases where alternative states are possible (e.g. in C, N and O) the state of lowest energy is not necessarily the state of maximum coupling of orbital motions. Thus ^1S states (which are *non-valent* states) for C and O are higher than both D and P states.

(2) The state of *maximum* multiplicity $(2S+1)$ among alternative states is the *lowest* state (Hund's rule, 1925). As the number $2S$ represents the minimum valency, this fact is of chemical significance. Thus nitrogen would be univalent in the states ^2D and ^2P, but possesses tervalency in the ground-state ^4S. Oxygen is non-valent in the states ^1D and ^1S, and bivalent in its ground-state ^3P. For states of the same multiplicity that of greatest L value is usually the lowest.

It is seen that the series of term-symbols of ground-states (as determined by experimental spectroscopy) is identical in Series II and III. This is clear proof that in the third shell ($= 3$), filled as regards s- and p-orbits as we pass from Na to Ar, the electrons take in succession the same orbital dispositions as in Series II. It is most important to realize the force of this purely physical proof of the analogy between elements in the two series, for, as will shortly appear, chemical evidence on this point is not conclusive. The actual electronic configurations of Ne and Ar are identical, as regards the outer shells, but Ar differs, at least potentially, from Ne in the fact that the former has in the outer shell no less than five vacant d-orbits, which could receive ten electrons. These d-orbits are, however, completely latent chemically, and we can only conclude that they lie at such a level above the p-orbits that p-electrons cannot undergo promotion to them in chemical action (cf. carbon, p. 115 and p. 155 below). The chemical inertness of argon, in spite of the possibility, presented by the presence of vacant d-orbits, of uncoupling and thus rendering chemically active any or all of its outer electrons, might well lead us to expect confidently that the vacant d-orbits in all the elements Na to Cl would also not be available. The limitations placed on the valencies of the elements H to Ne, summarized by the phrase 'octet rule', would then operate also in the series Na to Cl. That the lowest levels of d-orbits are actually high above the levels of the ground-states for this series is confirmed from spectroscopy.

With the object of testing this expectation we may consider the compounds SiF_6^-, PF_5, PCl_6^- (in solid PCl_5, p. 48) and SF_6. The fluorides PF_5 and SF_6 are amongst the most stable known compounds. As has been proved already (pp. 47, 72) all the linkages in each of the four compounds are equivalent in all respects, a fact losing some of its cogency in that the operation of resonance does not require us in such cases *necessarily* to exhibit all bonds as equivalent in any *one* bond-diagram. We may preserve a valency group of eight electrons in the central atom in at least two ways: (a) by reducing some of the bonds to single-electron links, (b) by reducing some of the bonds to ionized links. The theory of quantum mechanics shows that although single-electron links are not impossible, they could bind in stable union only certain atoms, of low atomic weight (p. 185). The assumption that one or more of the links is ionized is more promising. For PF_5 and SF_6 we could write *as one member of a resonance group*:

$$\begin{array}{ccc} & F^- & \\ F\diagdown & | & \diagup F \\ & P^+ & \\ F\diagup & & \diagdown F \end{array} \qquad\qquad \begin{array}{ccc} & F^- & \\ F\diagdown & | & \diagup F \\ & S^+_+ & \\ F\diagup & F^- & \diagdown F \end{array}$$

It must be clearly understood that the effect of this modification in bonding would be to increase the *polarity* of all the bonds *equally*, and not to cause actual ionization, which is known not to take place with these fluorides. Such an increase of electric moment cannot unfortunately be brought to experimental test, as the resultant moment of both PF_5 and SF_6 is zero from symmetry. Such a formulation is, however, not feasible for the ions SiF_6^- and PCl_6^-, for these compounds cannot be formed, or be regarded as formed, from neutral atoms. In the formation of

$SiF_6^=$ in the usual way from SiF_4 and F^-, no positive charge is present on the Si atom to bind the F atoms, and ionic linkages between SiF_4 and F^- are impossible. In view of this restriction it seems gratuitous to accept the formulation for SF_6 and PF_5 alone.

Very great interest attaches to the recent X-ray examination of crystalline PCl_5 and PBr_5. While the former is built of the ions PCl_4^+ and PCl_6^- (mutually disposed in a CsCl lattice), the bromide is formed of the ions PBr_4^+ and Br^-. In PCl_4^+ and PCl_6^- the P, Cl distances are respectively 1·98 and 2·06; in PCl_3, P, Cl = 2·04. In the PBr_5 lattice Br^- is about twice as far from the P atom as the four bonded Br atoms.* It may well be that the difference in structure is determined by the relatively large size of Br, which renders the ion PBr_6^- unstable.

Below are tabulated the classical formulae for certain oxygen compounds of Series III elements, contrasted with 'octet formulae', i.e. electronic formulae which, by the use of co-ionic bonds, preserve the valency group of eight electrons.

Sulphur: SO_2, SO_3, $SOCl_2$, SO_2Cl_2, $(SO_4)^=$

Classical diagrams   'Octet' diagrams

Phosphorus: $POCl_3$, POF_3, $(PO_4)^{\equiv}$

Chlorine: $(ClO_4)^-$

In SO_2 and SO_3 the resonance of the octet forms will bring

* Powell, Clark and Wells, *Nature*, **145**, 149, 971 (1940) and *J. Chem. Soc.* 1942, p. 642.

them, in respect to electric moment, bond-lengths and interbond angles, close to the classical formulae. In $SOCl_2$ and SO_2Cl_2 resonance, if not completely excluded, is unimportant. Hence in these molecules we cannot contemplate ionized linkages, as such would entail actual ionization of the compounds in solvents. In both chlorides the S, O distances are equal, and indubitably indicate at least double bonds between S and O. While co-ionic links are possible between S and O, they are not admissible between S and Cl. All indications seem here to support the classical formulae, which demand a valency group in S as large as twelve electrons. In $POCl_3$ and POF_3, as in the oxychlorides of sulphur, the bond-distance P, O demands double bonding between P and O, but owing to the possibility of resonance between the three Cl atoms, one of these could be held by an ionized link, and the octet be maintained in phosphorus.

Table 24 shows bond-distances in the commonest XO_4 anions, which adopt the form of a regular tetrahedron.

Table 24. *X, O distances in XO_4 ions*

	SiO_4^{4-}	PO_4^{3-}	SO_4^{2-}	ClO_4^-
X, O in XO_4^{n-}	1·61	1·55	1·51	1·46*
Maximum X, O	1·63	1·62	1·63	1·70
Contraction	Small	0·07	0·12	0·24
X, O (Table 6)	1·90	1·84	1·78	1·73

* Mean value: Gottfried and Schusterius, *Z. Krist.* 84, 65 (1932).

The maximum X, O lengths observed, which occur in cristobalite (p. 40), P_4O_6 (p. 74), polymerized ('asbestos') SO_3†, and Cl_2O (p. 89), offer the best evidence available for the lengths X—O. The deficit between these and the radius sums from Table 6 is due to difference of electro-negativities, which certainly lie in the order Si < P < S < Cl < O (see p. 45). Modernized 'classical' formulae for the ions XO_4^{n-} demand resonance between single and double bonds, the number of the latter increasing from zero in SiO_4^{4-} to three in ClO_4^-, which is seen to be the order of increasing contraction in Table 24. It may well be the conclusion reached from this discussion of the valencies

† Westrik and MacGillavry, *Acta Cryst.* 7, 764 (1954).

exerted by certain elements in Series III that it would be at present realistic to transcribe the unchanged classical formulae into electronic style.

Table 25. SERIES IV—Sc to Ni

	Configuration	Ground-state	$2S=$ minimum valency	Maximum obs. valency
Sc	$3d\ \ 4s^2$	^2D	1	3 (constant)
Ti	$3d^2\ 4s^2$	^3F	2	4
V	$3d^3\ 4s^2$	^4F	3	5
Cr	$3d^5\ 4s$	^7S	6	6
Mn	$3d^5\ 4s^2$	^6S	5	7
Fe	$3d^6\ 4s^2$	^5D	4	6 (unstable, in ferrates)
Co	$3d^7\ 4s^2$	^4F	3	3
Ni	$3d^8\ 4s^2$	^3F	2	2 (3?)
Cu	$3d^{10}\ 4s$	^2S	1	2

Turning now to Series IV, which begins with K, it is seen that the ground-states of this element and the next, Ca, are identical respectively with those of Na and Mg, and Li and Be. In respect to valency these three pairs of elements are indeed to be regarded as completely analogous. The succeeding elements from Sc to Ni show ground-states of high L and S values quite at variance with those of the latter six elements in the first two series. The ground-states of K and Ca (^2S and ^1S respectively) prove that their valency groups contain respectively one and two electrons in the $4s$-orbit. If, to produce Sc, the electron added to the Ca system adopted a $4p$-orbit, then the ground-state of Sc must be a P state (p. 149). As actually a D state is found this electron must occupy a d-orbit (p. 146), and there can be no doubt that this is a $3d$-orbit. Fig. 35 shows first the term-scheme of the (arc) spectrum of Al, i.e. the scheme of the spectrum arising from the excitation of one electron in the neutral atom; and, secondly, the scheme of the (spark) spectrum arising from the excitation of Al^{++}. It will be clear that the spectrum of Al (P ground-state) depends on the excitation of the $3p$-electron in the group $3s^2\ 3p$, and the spectrum of Al^{++} (S ground-state) on the excitation of the $3s$-electron remaining after the other $3s$-electron and the $3p$-electron

* Cf. Lister and Sutton, *Trans. Faraday Soc.* **35**, 495 (1939).

have been removed by ionization. Fig. 36 shows the term-scheme for the (spark) spectra of Sc^{++} and Ti^{+++}. Both exhibit a very similar form, with D ground-state, but are quite different from the scheme of Al^{++}. Both spectra clearly arise from the excitation of a $3d$-electron, remaining when two electrons ($4s^2$) have been

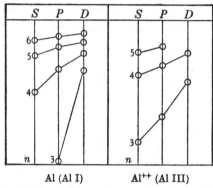

Al (Al I)   Al^{++} (Al III)

Fig. 35. Term-schemes for neutral and ionized Al.
Ground-states: Al; P, (Ne) $3s^2 3p$. Al^{++}; S, (Ne) $3s$.

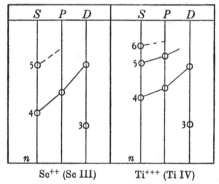

Sc^{++} (Sc III)   Ti^{+++} (Ti IV)

Fig. 36. Term-schemes for Sc^{++} and Ti^{+++}.
Ground-states: Sc^{++}; D, (A) $3d$. Ti^{+++}; D, (A) $3d$.

removed by ionization from the neutral Sc atom, and three electrons ($4s^2$ and $3d$) from the Ti atom; in each case the configuration (A) $3d$ remains. These term-schemes in fact prove conclusively that $3d$-orbits are occupied in both Sc and Ti. Fig. 37 shows in another way the relation of the term-schemes of the isoelectronic systems K (K I), Ca$^+$ (Ca II), Sc^{++} (Sc III), Ti^{+++} (Ti IV) and V^{++++} (V V), each of which has nineteen electrons, i.e. (argon) + 1. The atomic number Z is plotted against the square root of the wave-number (Appendix): cf. Moseley's law, p. 116. The parallelism of the graphs connecting $4s$, $4p$ and $4d$ states will be noted, while the position of the $3d$ graph proves that the active electron in Sc^{++}, Ti^{+++} and V^{++++} is a $3d$-electron; in K and Ca^+ on the contrary a $4s$-electron is active. The fact emerges that $3d$-orbits are so high above $3p$-orbits that they must be regarded as constituents of the fourth shell rather than of the third. The ground-states of the elements Sc to Ni are in complete agreement with the conception

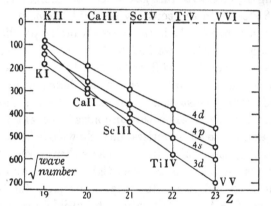

Fig. 37. Moseley's law for spectra of K, Ca^+, Sc^{++}, Ti^{+++} and V^{++++}.
(After White, *Phys. Rev.* **33**, 539 (1929).)

that in this sub-series the five possible d-orbits fill up with their complement of ten electrons, although, as will be seen (Table 25), the filling does not occur quite regularly, and requires nine elements for completion.

Comparison of the observed maximum valency of the elements Sc to Mn (inclusive) with the electronic configuration shows that the maximum valencies demand the active use of all the d- and s-electrons, some promotion being required in all elements except Cr to uncouple the two $4s$-electrons. Valencies therefore rise steadily to a maximum of 7 (in Mn). As we proceed further the maximum valency abruptly diminishes as the d-group becomes nearly complete and its electrons firmly bound. In the very unstable and little-known ferrates (anion FeO_4^-) in which Fe is sexavalent we see the last instance of the active use of all or nearly all the d-electrons. To form the cupric ion Cu^{++} one d-electron and the sole s-electron must be removed from the atom. It is therefore probable that Ni^{++} and Co^{++}, the common ions of these metals, are formed in the same way. Ni shows an unstable tervalency in a very few compounds, and Cu in none.

The simple, i.e. elementary, ions (cations or anions) produced by chemical action in aqueous solution, of all elements up to and including Sc, possess a 'rare gas' electronic configuration, i.e. are in an ^1S state. On the contrary no stable elementary ion from Ti to Cu inclusive can be in this state. Running parallel with this distinction in structure we find that (a) all simple ions up to Sc inclusive are colourless, while those from Ti to Cu are all coloured; (b) the colourless simple ions are all *diamagnetic*, while the coloured are all *paramagnetic*. Absence of colour in the ions up to Sc is the rule even when the un-ionized element is coloured: S^-, Br^- and I^- are all colourless. As is shown by the inertness and high ionizing potential of the rare gases themselves, electrons in the rare gas configuration are very tightly bound, and the energy quantum required to excite them is too large for the related frequency ($E = h\nu$) to fall within the visible spectrum. Paramagnetism, due to the presence of uncompensated magnetic shells, can only arise in states of multiplicity greater than unity. After Cu no more coloured or paramagnetic ions are found to be produced from the elements Zn to Br inclusive. In the eight elements Cu to Kr the $4s$- and $4p$-orbits become tenanted step by step as in Series II and III.

A substance is said to be *paramagnetic* when its presence in a magnetic field causes an enhancement of that field (permeability > 1). The property is directly attributable to the existence of un-compensated magnetic shells in the atom of the paramagnetic, which may in general be produced by both orbital and spin motions of electrons. In an external magnetic field the atomic shells are oriented to an extent dependent inversely upon the temperature (cf. electric moment, p. 35). If μ_0 is the value of the fixed g.-atomic magnetic moment the average moment $\bar{\mu}$ of an assembly of paramagnetic atoms (1 g. atom) in a field H is

$$\bar{\mu} = \frac{\mu_0^2 H}{3RT},$$

a result recalling that for electric moment (p. 37). A solid, close-packed assembly of paramagnetic atoms may exhibit the phenomena of *ferromagnetism*, all the circumstances of which are not yet fully understood.

Picturing an atomic electron classically, as a point-charge moving in a defined orbit, we must infer that, by classical theory, the orbit forms a magnetic shell of moment $M = AI/c$, where A is the area enclosed by the orbit, I the equivalent current and c the velocity of light. The equivalent current $I = e\omega/2\pi$, where ω is the angular velocity of the electron. From the quantum theory, as used by Bohr, the lowest angular momentum possible to an electron is $h/2\pi = mr^2\omega$, where m is the electronic mass and r the radius of the orbit. Hence the magnetic shell associated with the orbital motion of an electron in its lowest state with angular momentum (p-orbit) is

$$M = \frac{e}{2mc} \cdot \frac{h}{2\pi} = 9 \cdot 18 \times 10^{-21} \text{ unit pole} \times \text{cm.}$$

This quantity, known as the *Bohr magneton*, clearly represents a natural unit of atomic magnetism.* The fact that magnetic moment must also be attributed to an electron by reason of its *spin* is illustrated by the example of nitric oxide NO (electronic formula, p. 127 and p. 190). The ground-state of this 'odd electron' molecule is exceptional in being a close doublet (cf. p. 149), spectroscopically described as $^2\Pi_{\frac{1}{2},\frac{3}{2}}$ (p. 191). In both states the 'odd' electron occupies a p-orbit (more precisely, a π-orbit; p. 180) and on this account is associated with a definite magnetic moment.

* Although the angular momentum deduced by quantum mechanics for a p-electron is $\sqrt{l(l+1)} \cdot h/2\pi = \sqrt{2} \cdot h/2\pi$, the *component* of the momentum in the direction of an external magnetic field is still $h/2\pi$, i.e. the axis of the momentum precesses round the field direction at the constant angle $\pi/4$. For discussion see Herzberg, *Atomic Spectra and Atomic Structure*, 1937, p. 101, or other full account of the Zeeman effect.

The lower state $(^2\Pi_{\frac{1}{2}})$ is however *diamagnetic*, and we must assume that the spin of the electron, opposed in direction to the orbital motion, is also associated with an *equal* moment. In the light of Bohr's original theory, in which a p-electron had angular momentum of $h/2\pi$, while to the spin only $\frac{1}{2}.h/2\pi = h/4\pi$ could be attributed, this result is unexpected. The upper state of the doublet $(^2\Pi_{\frac{3}{2}})$ is paramagnetic, giving to NO a resultant molecular magnetic moment* (see further, p. 191). In this state the orbital and spin magnetic moments reinforce each other. The existence of spin magnetic moment and its relation to orbital magnetic moment has been proved experimentally in a study of the effect of magnetic fields upon spectra (the Zeeman effect), and theoretically by Dirac in an application of relativistic principles. The calculation of resultant magnetic moments arising from the coupling of both orbital and spin angular momenta proves to be complicated, and cannot yet be said to be completely explored.† When,

Table 26

S	0	$\frac{1}{2}$	1	$\frac{3}{2}$	2	$\frac{5}{2}$
Atomic moment $2\sqrt{S(S+1)}$	0	1·73	2·83	3·88	4·90	5·91

Table 27. *Ions of transitional elements*

	Observed atomic moment (magnetons)	S (from Table 26)	Valency of ion (z)	$z + 2S$
V^{++++} (blue)	1·72	$\frac{1}{2}$	4	5
V^{+++} (green)	2·70	1	3	5
V^{++} (violet), Cr^{+++} (green)	3·85	$\frac{3}{2}$	2, 3	5, 6
Cr^{++} (blue), Mn^{+++} (red)	4·85	2	2, 3	6, 7
Mn^{++} (pink), Fe^{+++} (yellow)	5·8	$\frac{5}{2}$	2, 3	7, (8)
Fe^{++} (green), Co^{+++} (red)	5·3	2 (?)	2, 3	6, (7)
Co^{++} (red, blue)	4·85	2	2	
Ni^{++} (green)	3·2	1 (?)	2	
Cu^{++} (blue)	1·9	$\frac{1}{2}$	2	
Cu$^+$, Zn^{++} (colourless)	0	0	1, 2	

* For a discussion of NO (and O$_2$) and the relation of magnetic moment to temperature see van Vleck, *Nature*, 119, 670 (1927) and p. 194.   † See footnote, p. 161.

however, the resultant magnetic moment arises solely from electron spin, with resultant spin number S, the calculated molecular magnetic moment is $2\sqrt{S(S+1)}$ magnetons (Table 26).

There is some reason for thinking that for hydrated ions, i.e. ions in solution, L may be effectively zero, and the atomic magnetic moment due entirely to electron-spin S. It is indeed seen from the above data for V, Cr, Mn and Fe that the maximum valency of the element = (valency of the ion + $2S$).* In regard to the colours of the ions it may be observed that hydration frequently causes a modification of tint, but rarely creates colour. Colour in compound ions (e.g. CrO_4^-, MnO_4^-) is not *necessarily* attributable to the structure of the central atom, but is likely to be closely dependent on its properties.

The fourth series terminates with Kr, the configuration of whose outer shell is $4s^2\,4p^6\,3d^{10}$. Now a completed fourth shell ($n = 4$) should theoretically comprise besides $4s$- and $4p$-electrons ($l = 0$ and 1, respectively) ten $4d$-electrons ($l = 2$) and no less than fourteen $4f$-electrons ($l = 3$). It is clear that as regards the occupancy of possible orbits Kr is much less 'complete' than A, which only lacks occupancy of the $3d$-orbits. The set of elements from K (Z 19) to Kr (Z 36) inclusive may both on chemical grounds and for reasons of atomic structure be subdivided as follows:

Set I		No. of elements	Type of orbit filled
Section 1	K—Ca	2	$4s$
2	Sc—Ni	8	$3d$
3	Cu—Kr	8	$4s$ and $4p$

If we examine the elements between Rb (Z 37) and Xe (Z 54) and then those between Cs (Z 55) and Rn (Z 86), we find that again on chemical grounds we are encouraged to discover similar sub-sections 1, 2 and 3 in both these later sets of elements, as follows (cf. periodic system, Table 2, p. 24).

* A compact account of atomic and molecular magnetism and its problems will be found in Stoner, *Magnetism* (Methuen, 1936), which contains a bibliography of larger works.

Set II	No. of elements	Type of orbit filled
Section 1 Rb—Sr	2	$5s$
2 Y—Pd	8	$4d$
3 Ag—Xe	8	$5s$ and $5p$

Set III	No. of elements	Type of orbit filled
Section 1 Cs—Ba	2	$6s$
2 La—Pt	8	$5d$
3 Au—Rn	8	$6s$ and $6p$

The equality in the membership of sub-sections in each of the three sets of elements, the close analogy in the chemistry of corresponding elements (see below), the exhibition of paramagnetic and coloured ions at the expected points in the succession, and finally the close similarity of the ground-states, all affirm a common cause for the type of succession in the three sets, which, as indicated in the third columns above, is regarded as the progressive filling of the appropriate d-orbits between the completion of the s- and p-orbits.

Notes on compounds of certain metals, included in Table 28, p. 163

SET II

Tc: Very little is at present known about this recently discovered element.

Ru: The highest fluoride of Ru is RuF_5, which on treatment with water undergoes hydrolysis and disproportionation to RuO_4 and lower oxides. RuO_4 closely resembles OsO_4, especially in volatility. This must be one of the few cases in which F does not excite the highest valency.

Rh: The highest fluoride of Rh is RhF_3, whose crystal structure is the same as those of PdF_3, CoF_3 and FeF_3 (see below). RhO_2 is most probably a peroxide, with the grouping —O—O—.

Pd: PdF_3 yields $Pd(OH)_2$ and O_2 upon hydrolysis with cold water.

SET III

Re: Re_2O_7 consists of volatile yellow crystals, m.p. 220°; it sublimes readily even at 150°. It is stable to 800°, and its vapour density corresponds to Re_2O_7.

Os: OsF_8, m.p. 34°, b.p. 47°, is stable at least to 400°. OsO_4 is very volatile. Its crystal structure shows molecules of the tetroxide with atoms of O tetrahedrally disposed round the central Os atom.

Ir: IrF_6 has m.p. 44° and b.p. 53°.

Table 28. *Valencies of the transitional elements*

Set I		Set II		Set III	
Sc	3	Y	3	La	3
Ti	4	Zr	4 (Chap. III)	Hf	4 HfCl$_4$
					(HfCl$_6^-$)
					HfO$_2$
V	5	Nb	5 NbCl$_5$	Ta	5 TaCl$_5$
			NbF$_5$		Ta$_2$O$_5$
			Nb$_2$O$_5$		
Cr	6	Mo	6 MoF$_6$	W	6 WF$_6$
			MoO$_3$		WCl$_6$
					WO$_3$
Mn	7	Tc	(7)	Re	7 ReCl$_7$
					Re$_2$O$_7$
					ReO$_4$
Fe	(6) 3	Ru	8 RuO$_4$	Os	8 OsF$_8$
					OsO$_4$
Co	3	Rh	3 (4?)	Ir	6 IrF$_6$
			RhF$_3$		
			(RhO$_2$)		
Ni	2	Pd	2 (3?)	Pt	4 PtCl$_4$, etc.
			PdCl$_2$		
			PdF$_3$		

In all the sets the valency falls abruptly after attaining its maximum, which reaches 8 in Sets II and III. The valencies of the first five elements in each set correspond exactly.

According to the work of Ebert* the crystal structures of the trifluorides of Rh, Pd, Fe and Co are unusual, in that the metallic atom lies at the centre of a triangular anti-prism, of which the six corners are occupied by halogen atoms. Every fluorine atom has two metal atoms as its equidistant and nearest neighbours (6:2 co-ordination). This interpretation has recently been critically re-examined.†

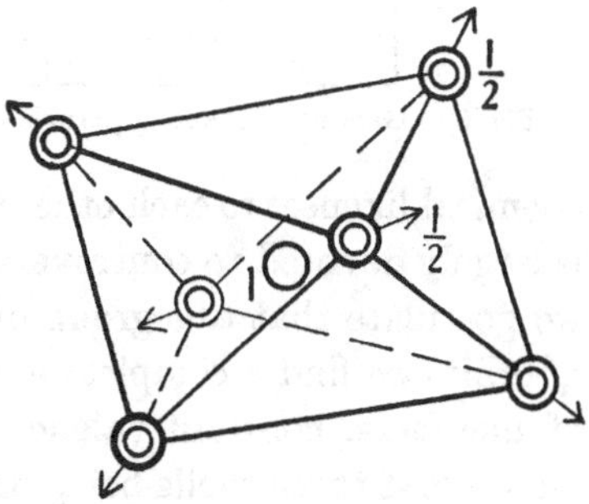

In order to bring to light the

Fig. 38. The structure of FeF$_3$.
O, Fe; ◎, F.

* *Z. anorg. Chem.* **196**, 395 (1931).
† Hepworth, Jack, Peacock & Westland, *Acta Cryst.* **10**, 63 (1957).

obvious analogies in the above three sets of elements a whole group of fourteen metals has been overlooked between La and Pt in section 2 of Set III. These elements between Ce (Z 58) and Lu (Z 71), inclusive, have become known as the *rare-earth group*. In many respects this group shows similarity to what might be shortly called the '*d*-orbit' sections of the Sets I, II and III. Where the ground-states are certainly known they show high L and S values: all the ions are paramagnetic, and most of them are coloured. The important point of difference is that the rare-earth metals show a nearly constant tervalency, and bear such close

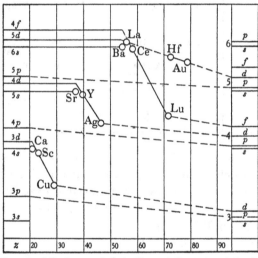

Fig. 39. Genesis of the long periods in the periodic system (after Herzberg).

chemical likeness to each other that the history of their discovery is largely devoted to controversies on methods of separation. If we postulate that this group owes its origin to the filling of the 4*f*-orbits we find a complete explanation of its position, number of members, constant valency and other properties. The completed first three shells being omitted, the configuration of La is $4s^2 4p^6 4d^{10}$ (4*f* absent); $5s^2 5p^6 5d 6s^2$, giving a ground-state ^2D like that of Sc. Except for the progressive filling of the seven 4*f*-orbits by fourteen electrons this configuration remains constant until Lu is reached, and the *valency* group remains always $5d\, 6s^2$, giving, with promotion of an s-electron, tervalency, as in Sc and La. The metal Hf (Z 72) contains a completed fourth shell, and its valency group becomes $5d^2 6s^2$, giving quadrivalency on promotion (cf. carbon), and definitely placing Hf in Group IV. It is interesting to notice that Ce, Pr and Nd also exhibit an (unstable) quadrivalency (p. 73), Ce^{4+} having the configuration of xenon. Fig. 39 is an attempt to show diagrammatically how the available atomic levels change their relative positions as atomic number Z increases *pari passu* with the addition of electrons.

'Artificial' elements

Of the elements between H and U, four, Technetium (43, Tc), Promethium (61, Pm), Astatine (85, At) and Francium (87, Fr), are known only as products of appropriate nuclear reactions, from which they emerge as radioactive isotopes. The half-lives of the best-known species are as follows:

Tc^{99} (Group VII A), 10^6 yr. Pm^{147} (rare-earth group), 2·5 yr.
At^{210} (Group VII B), 8·5 yr. Fr^{223} (Group I A), 20 min.

It has been confirmed that Tc is closely similar to Re, and that At has some of the properties expected of a halogen, such as forming the anion At^-.

The most remarkable advance made by modern nuclear chemistry is the production of at least six 'transuranic' elements, that is elements surpassing the apparent 'natural' limit of atomic number at 92. These elements, which are all radioactive metals, have been named and allotted symbols as follows:

Neptunium, Np 93; Plutonium, Pu 94; Americium, Am 95; Curium, Cm 96; Berkelium, Bk 97; Californium, Cf 98.

The chemistry of the first four is summarized below as regards their valencies; U is also included, and the most stable species are denoted by heavy type.

A steadily increasing tendency towards greater stability of the lower valencies is evident until Cm is reached, which shows little tendency to leave the tervalent state. In relation to the periodic system two interpretations of the transuranic series are plausible: (1) that it is a part of a fourth transition series, and should bear some resemblance to the corresponding members of the third series, La 57 to Au 79; (2) that it is part of a series corresponding to the rare-earth group (or *lanthanides*), in which 5f orbits play the part of 4f orbits in the latter. If there were an exact

Transuranic elements

Valency	U	Np	Pu	Am	Cm
III	U^{3+}, 5f^3	Np^{3+}	Pu^{3+}, (5f^5)	Am^{3+}, (5f^6)	Cm^{3+}, (5f^7)
IV	U^{4+}, 5f^2	Np^{4+}	Pu^{4+}	Am^{4+} (in AmO$_2$)	Cm^{4+} (in CmF$_4$ and CmO$_2$)
V	UO$_2^+$	NpO$_2^+$	PuO$_2^+$	AmO$_2^+$	—
VI	UO$_2^{2+}$ UO$_3$	NpO$_2^{2+}$	PuO$_2^{2+}$	AmO$_2^{2+}$	—

$\updownarrow$
X Indicates disproportionation.

correspondence then the '5f' series should begin with Th, and U would be the third member. There is, however, a strong similarity between Th and Hf, and no evidence that 5f orbits are occupied in the normal states of either Th or Pa. On the contrary, as indicated in the summary, evidence that 5f orbits are occupied in several states of U is strong and generally accepted. Indications about 5f orbits placed in parentheses in the summary are less certain but very probable. The term 'actinide' corresponding to 'lanthanide' is at present regarded as justified for U and the succeeding heavy elements.

Chapter VI

SOME CURRENT TOPICS IN VALENCY THEORY

SECTION I. CO-ORDINATION COMPOUNDS

It has long been known that very stable 'molecular compounds' can often be generated by the union of simpler bodies, in which the atoms seem already to be exerting their maximum valencies, and some examples have already been mentioned. The largest number of instances is found in the class of *complex ions*, which are sometimes defined as resulting from the union of a neutral molecule with an anion or kation. The distinction between 'compound ion' and 'complex ion' is largely arbitrary and originated historically. It was formerly the custom to formulate a molecular compound by writing the generators side by side, e.g. $2KI.HgI_2$; $2KCl.MgCl_2$, all compounds such as these being called '*double salts*'. A cryoscopic examination of these two 'double salts' in solution disclosed a fundamental difference, for while both yield K^+ ion, the former yields a total of only three, but the latter, a total of seven, ions per mol. The formulation of the mercuric compound was therefore changed to $K_2[HgI_4]$, and the ion HgI_4^-, formally resulting from the reaction

$$HgI_2 + 2I^- = HgI_4^-,$$

was termed a *complex* ion, and the corresponding salts *complex salts*. From the standpoint indicated in this example ions such as SO_4^-, NO_3^-, etc., although 'compound', were not regarded as 'complex'; they did not seem to raise the perplexing problems of valency obviously involved in the complex ion.

In the very numerous complex ions containing NH_3 units (e.g. metallic ammines) and CN^- ions (e.g. ferrocyanides) it was formerly possible to urge that chemical combination within the complex was due to a rise in effective valency of nitrogen from 3 to 5, as in NH_4Cl, which may itself be regarded as $NH_3.HCl$. In complexes involving water or OH^- oxygen could be assumed quadrivalent, and so on. Such a 'classical' explanation of complex substances is excluded in the important and numerous class of complex fluorides, examples of which are tabulated below.

Complex fluorides

(General mode of preparation $MF_n + mKF \rightarrow K_m \cdot MF_{n+m}$.)

General formula	Element M	Generating fluoride
RMF_4	B (tervalent)	BF_3
R_2MF_4	Be, Zn, Cd, Hg (bivalent)	$(M)F_2$
R_2MF_6	Si, Ti, Sn, Pb, Pt (quadrivalent)	$(M)F_4$
R_3MF_6	Al (tervalent)	AlF_3
R_2MF_7	Ta, Nb (quinquevalent)	$(M)F_5$
R_3MF_7	Zr (quadrivalent)	ZrF_4
R_3MF_8	Ta (quinquevalent)	TaF_5

(R = alkali metal or NH_4.)

The table comprises only a small selection of the known complex fluorides, for under the heading M only those elements have been included that are exerting, in the generating fluoride, their highest known (classical) valency. The stereochemistry of the

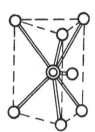

Fig. 40. The structure of $(NbF_7)^=$.
⊚, Nb; ○, F.

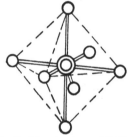

Fig. 41. The structure of $(ZrF_7)^{3-}$.
⊚, Zr; ○, F.

complex ions of this class is readily discovered by X-ray analysis, by the use of which it is found that MF_4 is tetrahedral and MF_6 octahedral. MF_7 is found to take alternative forms: $NbF_7^=$ * being based on the trigonal prism, with the additional F atom at a face-centre (fig. 40), and ZrF_7^{3-} (and $NbOF_6^{3-}$)† on the octahedron, with the additional F again at a face-centre: an axis of trigonal symmetry passes through the central Zr and the face-centred F† (fig. 41).

* Hoard, *J. Amer. Chem. Soc.* **61**, 1252 (1939).
† Hoard and Williams, *J. Amer. Chem. Soc.* **64**, 1139 (1942).
‡ Hampson and Pauling, *J. Amer. Chem. Soc.* **60**, 2702 (1938).

The only 8-co-ordinated complex yet completely analysed is the molybdenocyanide ion $Mo(CN)_8^{4-}$. The configuration of this ion is obtained by erecting upon each of the four faces of a regular tetrahedron a low regular pyramid, thus forming a triakis-tetrahedron, with twelve faces and eight apices. The directions of the eight bonds are then found by joining the centre of the original tetrahedron, where is placed the Mo atom, to each apex. The axis of the linear group CN lies along the bond direction, and the centre of the group at the apex (fig. 42).* TaF_8^{3-} appears to take the form of a square anti-prism.† In every example of a complex fluoride ion so examined the M, F distances have been found exactly equal, and therefore all bonds equivalent.

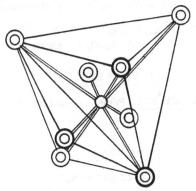

Fig. 42. The structure of the ion $Mo(CN)_8^{4-}$.
◎, CN; ○, Mo.

The mode of preparation, stability and chemical properties of typical complexes containing metals may be illustrated by a short account, clarified by tabulation, of some long-known examples containing cobalt. In most circumstances cobaltous salts (Co^{++}) are oxidized to the tervalent (Co^{+++}) state only with difficulty, and then yield very unstable compounds. If, however, NH_3 and NH_4Cl are dissolved with the cobaltous salt, oxygen is rapidly absorbed even from a stream of air, with the production of several stable compounds.

* Hoard and Nordsieck, *J. Amer. Chem. Soc.* **61**, 2853 (1939).

† Hoard (Marchi, Fernelius and McReynolds, *J. Amer. Chem. Soc.* **65**, 330 footnote (1943)).

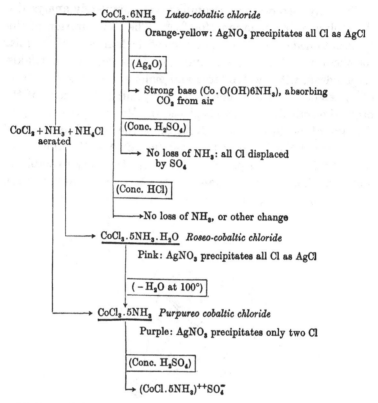

Failure of chlorine apparently present as chloride to respond to the presence of $AgNO_3$ was also noted in the hydrated chlorides of Cr (p. 101). Werner was the first to perceive that the reactions of the Co compounds above, and the numerous others like them, demand such a formulation as follows:

$$[Co, 6NH_3]^{+++} 3Cl^-, \quad [Co, 5NH_3 . H_2O]^{+++} 3Cl^-,$$
$$[Co, 5NH_3 . Cl]^{++} 2Cl^-,$$

and to recognize the very strong bonding to the central metallic atom of the six groups enclosed within the bracket. He introduced the description 'co-ordination' for this bonding, and the term 'co-ordination number' for the total number of co-ordinate bonds. In crystallography the terms 'co-ordination' and 'co-ordination number' have long been used more broadly, and refer only to the number of nearest neighbours, without consideration of the nature of the linking. Thus crystallographically the octahedral arrangement of six chlorine ions round any sodium ion in NaCl, and the similar disposition of six NH_3 round Co^{+++} in Co complexes, are both equally termed examples of 6-co-ordination.

It is now known that a very wide range of groups can be so bonded. The following may be mentioned without exhausting the list: NH_3, amines, pyridine; water, oxygen and NO; alkyl sulphides R_2S, phosphines R_3P, arsines R_3As; the ions CN^-, $(hal.)^-$, OH^-, NO_2^-. All these function as one co-ordinating group. Diamines, such as $NH_2CH_2CH_2NH_2$, and bivalent ions, such as oxalate $C_2O_4^-$, can replace two groups of the former class, while triamines can occupy three places in the complex. As examples of more complicated groupings we may take

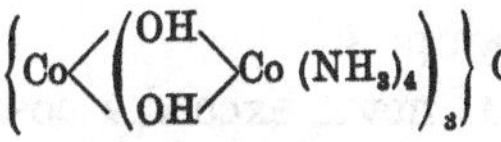

$$\left\{ Co\!\!\left<\!\!\binom{OH}{OH}\!\!Co\,(NH_3)_4\!\right)_{\!3}\right\} Cl_6 \quad \text{(Werner)}$$

and $\quad\left\{\!\!\left(SO_2\!\!\bigg<^{NH}_{NH}\!\!\right)_{\!2}\!\!Rh\!\!\bigg<^{OH_2}_{OH_2}\right\}^{\!-} Na \quad$ (Mann, *J. Chem. Soc.* 1933, p. 412).

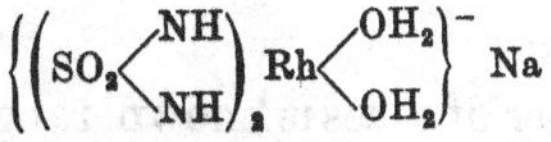

(Sulphamide, $SO_2(NH_2)_2$, neutralizes aqueous sodium carbonate, giving the salt $SO_2(NHNa)_2$.) In the latter example we find a typical illustration of the stereoisomerism that may arise in complex substances. The assumption of an octahedral arrangement around the metal atom (justified in numerous instances by direct X-ray examination) leads to the possibility of two isomeric dispositions (fig. 43):

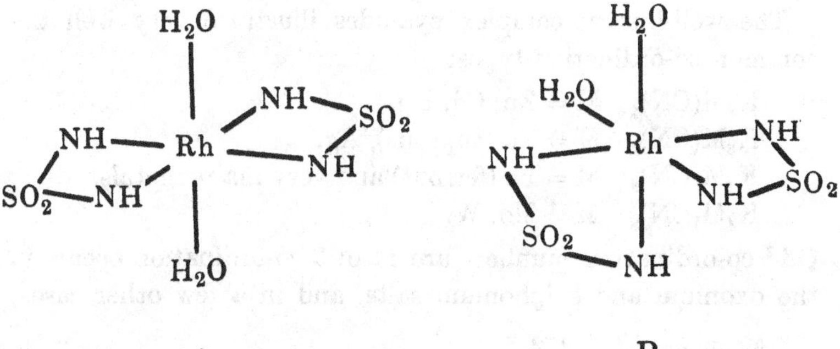

A   B

Fig. 43.

Of these B is dissymmetric and its successful optical resolution*
confirms the proposed arrangement (cf. pp. 34, 35).

Some examples may now be tabulated to indicate the maximum
co-ordination numbers commonly found and their distribution
among the elements.

4-co-ordination

    Li In various organic complexes.†

    Be BeF_4^-, and in the benzoyl-pyruvic derivative shown to
        be optically active.‡

    B BF_4^-, $BF_3.NH_3$, $BH_3.CO$, etc.

    C (Normal valency).

    N NH_4^+, etc.

    O Oxides such as ZnO and BeO (pp. 40, 55); ice (p. 223);
        basic Be acetate (p. 56).

    F No known examples.

(Elements in Series II never exceed 4-co-ordination.)

6-co-ordination

A very large number of cases is known, ranging upwards in complexity from the complex fluorides, e.g. SiF_6^-. The transitional
sets of metals, Cr to Ni, Mo to Pd and W to Pt, very frequently
play the part of the central metallic atom. Many metals exhibit
both 4- and 6-co-ordination, e.g. $K_2Ni(CN)_4$ and $K_4Ni(NO_2)_6$.

Higher co-ordination

The ions TaF_7^{2-}, TaF_8^{3-}, ZrF_7^{3-} and $Mo(CN)_8^{4-}$ are among the
relatively rare examples of co-ordination higher than 6.

The well-known complex cyanides illustrate very well the
common co-ordination types:

    $K_2M(CN)_4$ M = Zn, Cd, Hg.

    $K_3M(CN)_4$ M = Cu (cuprous), Ag.

    $K_4M(CN)_6$ M = Fe (ferrous) and very many metals.

    $K_4M(CN)_8$ M = Mo, W.

Odd co-ordination numbers are rare: 3-co-ordination occurs in
the oxonium and sulphonium salts, and in a few other cases,

---

  * Mann, *loc. cit.* p. 172.

  † Morgan and Burstall, *Inorganic Chemistry* (1936), pp. 61, 62.

  ‡ Mills and Gotts, *J. Chem. Soc.* 1926, p. 3121.

such as $KHgI_3$ and $K_2Ag(CN)_3$. 5-co-ordination seems only to occur in the carbonyls (see later), and 7-co-ordination has already been exemplified above.

All indications are that the bonds between central atoms and co-ordinated groups (or *ligands*) reach the stability of those in organic compounds. The similarity of some co-ordination complexes to ordinary organic molecules is well shown by the following examples of 4-co-ordinated Pd (palladous) compounds:*

Phosphine compounds. $(R_3P)_2PdCl_2$,

R = *n*-propyl to *n*-amyl; m.p. all under 100°.

Arsine compounds. $(R_3As)_2PdCl_2$,

R = *n*-amyl; m.p. 10° (liquid at room temperature).

Alkyl sulphide compounds. $(R_2S)_2PdCl_2$,

R = ethyl to iso-butyl; m.p. 32°–97°.

All these metallic compounds are freely soluble in organic solvents, and show no tendency to ionize. By demonstrating the potential optical activity of numerous co-ordination complexes Werner further confirmed the strength of bonds within the complex (p. 34). It has been proved† that the Pt compound

$$\begin{array}{c} NH_3 \\ | \\ C_2H_4-S^+ \\ | \\ C_2H_4 . NH_3^+Cl^- \end{array} \!\!\!\!\!\!\!\!\!\!\! >Pt^-(Cl)_4$$

is resolvable into optical isomers. Dissymmetry can only arise in this case if the S atom is 3-bonded (as in the active sulphonium salts) and, further, the bond is necessarily of the co-ionic type.

The accepted view at the present is that the co-ordinated groups form co-ionic links which utilize *vacant* orbits in the central metal atom or ion. This idea, supported by the fact that nearly all known ligands contain at least one free electron-pair‡ (p. 112), and by the limit of 4-co-ordination in Series II, frequently entails the reception by the central metal of a large negative formal charge: in $Fe(CN)_6^{4-}$ and $Mo(CN)_8^{4-}$ we must

---

* Mann and Purdie, *J. Chem. Soc.* 1935, p. 1549.

† Mann, *J. Chem. Soc.* 1930, p. 1745.

‡ C_2H_4 is an apparent exception (see Chatt, *J. Chem. Soc.* 1953, p. 2939).

assume no less than four negative charges upon the iron or molybdenum atom. It might be objected that the necessity of such charges, so repugnant to the electropositive tendencies of all metals, invalidates the theory; but this objection may not be serious, for, as in the case of carbon monoxide, $C^-\!\!\equiv\!\!O^+$, the formal polarity may be almost completely offset by the 'natural' polarity. It may be asserted that the postulate of co-ionic links has shown such promise in explaining and predicting the chemistry, stereochemistry, and particularly the magnetic properties, of complexes, that the idea cannot be fundamentally wrong.

The volatile metallic *carbonyls*, exemplified by the well-known Ni compound $Ni(CO)_4$, are closely related to the co-ordination compounds involving metallic ions. They may be obtained by the action of carbon monoxide, usually at high pressure and at 200–300°, directly upon the finely-divided metal, or upon compounds, especially sulphides or iodides in the presence of agents (e.g. Cu or Ag) to remove sulphur or iodine.

The metal carbonyls

M	$M(CO)_6$ (octahedral)	M	$M(CO)_5$ (trigonal bipyramid)	M	$M(CO)_4$ (tetrahedral)
Cr, 24	B.p. 420°	Fe, 26	M.p. −23°, yellow	Ni, 28	M.p. −25° B.p. 43°
Mo, 42	Sublimes 40°	Ru, 44	M.p. −22°	Pd, 46	—
W, 74	Sublimes 50°	Os, 76	M.p. −15°	Pt, 78	—
M	$M_2(CO)_{10}$	M	$M_2(CO)_8$	—	—
Mn, 25	M.p. 154°, yellow*	Co, 27	M.p. 51°, red	—	—
Tc, 43	—	Rh, 45	M.p. 76°, orange	—	—
Re, 73	M.p. 177°	Ir, 77	Green	—	—

(The carbonyls are colourless where a colour is not stated.)

* Brimm, Lynch and Sesny, *J. Amer. Chem. Soc.* **76**, 3831 (1954).

As will be seen from the table, metals of *even* atomic number yield carbonyls which are monometallic, and, excepting $Fe(CO)_5$, colourless; while from metals of *odd* atomic number bimetallic and usually coloured carbonyls are produced. In all examples yet examined by diffraction methods the metal is linked directly to carbon, and the C, O distance (1·15A.) in the monometallic carbonyls is close to that in free CO (1·13A.). If we postulate that the metal-carbon link is single and that its electron pair arises from the lone pair on carbon in carbon monoxide (see p. 123) then in the monometallic carbonyls the metal is formally zero-valent (since none of its electrons is used in bonding), and, moreover, the composition of the carbonyls is such that the number of electrons donated by the ligands raises the total electronic content of the metal atom (sometimes termed the *effective atomic number*, E.A.N.) to that of the next ensuing inert gas: for example

$$\text{Cr(CO)}_6: \ 24 + 12 = 36 \ (\text{Kr})$$
$$\text{Ni(CO)}_4: \ 28 + \ 8 = 36$$
$$\text{Ru(CO)}_5: \ 44 + 10 = 54 \ (\text{Xe})$$
$$\text{Os(CO)}_5: \ 76 + 10 = 86 \ (\text{Rn})$$

The vacant orbits, available to the ligands, in the atoms of common metals forming carbonyls are shown in the table below:

	Configuration (see also Table 25)		Vacant orbits and their electronic capacity			Possible no. of co-ordinate links	Shape
Cr	$3d^5$	$4s$	$3d^5$	$4s$	$4p^6$ (12)	6	⎫
Mo	$4d^5$	$5s$	$4d^5$	$5s$	$5p^6$ (12)	6	⎬ Octahedral
W	$5d^4$	$6s^2$	$5d^6$	—	$6p^6$ (12)	6	⎭
Fe	$3d^6$	$4s^2$	$3d^4$	—	$4p^6$ (10)	5	Bi-pyramidal (cf. PF$_5$, fig. 13)
Co	$3d^7$	$4s^2$	$3d^2$	—	$4p^6$ (9)	9 (from 2 atoms)	⎫
Ni	$3d^8$	$4s^2$	$3d^2$	—	$4p^6$ (8)	4	⎬ Tetrahedral
Cu	$3d^{10}$	$4s$	—	$4s$	$4p^6$	No carbonyl	⎭

At present the structure of only one bimetallic carbonyl is known in detail, that of $\text{Fe}_2(\text{CO})_9$ which can be produced from Fe(CO)_5 by photochemical action. In this compound the metal atoms lie symmetrically on the common axis of three equilateral triangles of CO groups:

$$(\text{CO})_3\text{Fe(CO)}_3\text{Fe(CO)}_3.$$

The six terminal groups are bound as in the monometallic carbonyls, but each of the carbon atoms in the central groups is bound to both metal atoms, and these groups closely resemble the ordinary ketonic $\diagup C{=}O$. With this constitution each metal atom attains the E.A.N. 35, and necessarily contains one uncoupled electron. An additional direct bond between the Fe atoms would secure for each an E.A.N. 36 as in $Fe(CO)_5$, and at the same time ensure the diamagnetism that the enneacarbonyl in fact exhibits. The Fe, Fe distance observed in the carbonyl is indeed almost equal to that in the metal (2·48 A.).

Although in the compound $BH_3.CO$ (pp. 60, 134) the B, C link is undoubtedly co-ionic and single, detailed structural examination of the monometallic carbonyls (mainly by electron diffraction) has not fully supported the postulate of single carbon-metal links upon which the principle of the effective atomic number is based. In all carbonyls the group metal —C—O is linear, but instead of at least 2·0 A. expected for single bonds, the metal-C distance in nickel and iron carbonyls is 1·82 A. and 1·84 A. respectively. A bond system which is the result of resonance between the linear forms $M^-\text{-}C{\equiv}O^+$ and $M{=}C{=}O$ recalls the system supposed to account for the structure of CO_2 (p. 129), and would not only explain the short M-C length but would tend to diminish the negative charge imposed on the metal by purely co-ionic links.

Orbits hybridized	Shape of AX	Examples
sp	Linear	$BeCl_2$, $HgCl_2$, $Ag(CN)_2^-$, (C_2H_2)
sp^2	Equilateral triangle with outer and central atoms co-planar	BF_3, $B(OH)_3$, (C_2H_4)
$\left.\begin{array}{l}sp^3\\sd^3\end{array}\right\}$	Regular tetrahedron	BF_4^-, CH_4, NH_4^+, $TiCl_4$
sp^2d	Square with donor and central atoms co-planar	$(Cu.4NH_3)^{++}$, $(Pd.4NH_3)^{++}$, $(Ni.4CN^-)^{--}$
sp^3d	Trigonal bi-pyramid	PF_5, $TaCl_5$, $NbBr_5$
sp^3d^2	Regular octahedron	SF_6, SiF_6^{--}, $(Co.6NH_3)^{+++}$

Very few of the very large number of known metallic complex ions have been studied in detail by diffraction methods, but at least for ligands such as NH_3 (and its derivatives) and the fluorine ion in complex fluorides only single co-ionic links appear to be possible, but even in this field the principle of effective atomic number is not generally applicable.

It would perhaps be appropriate to summarize as above the relations between the n bond-forming orbits hybridized in the central atom of a compound or complex ion and the stereochemistry of the n single links (co-ionic or covalent) formed by the central atom.

Salt hydrates and co-ordination

Co-ordination of water round the cation of a crystal will often result in the cation and anion becoming comparable in size. Thus although Al^{+++}, Be^{++} and Mg^{++} are very small (0·57, 0·34 and 0·78 respectively), $Al.6H_2O^{+++}$, $Be.4H_3O^{++}$ and $Mg.6H_2O^{++}$ are at least as large as the SO_4^- ion. This approach in size between cation and anion allows a simpler and more symmetrical structure to be produced than would be possible with the anhydrous salt. Thus the alums $R.Al.(SO_4)_2.12H_2O$ all crystallize in the cubic system;* $BeSO_4.4H_2O$ has the CsCl structure only possible when the cation and anion are nearly equal in size; $MgCl_2.6H_2O$ has an only slightly distorted CaF_2 structure.

* Cf. Lipson, *Proc. Roy. Soc.* A, **151**, 347 (1935).

SECTION II. THE ENERGY OF LINKS, MULTIPLE BONDS, AND THE METHOD OF MOLECULAR ORBITS

In preceding chapters the implied cause of bond formation was assumed to be that, by the formation of 'common' orbits, the bonded atoms were enabled to fill orbits, partially empty before union, with a complement of electrons; and a definition of valency was based upon this conception. From a classical viewpoint the genesis of a chemical bond lies in the energy released in its formation from the constituent atoms, so that the resulting molecule remains in a trough of energy from which its constituent atoms can only emerge by the input of its energy of formation. It may therefore be assumed empirically that in the overlapping of orbitals to form ultimately a single bond (p. 114 and fig. 30) energy is continuously released as the overlapping extends with decreasing interatomic distance. We may reasonably expect that a study of this relation between the formation of 'common' orbits and the release of energy will reveal the source of the energy, and so supply a major defect in classical chemical theory.

Release of energy from a system of two mutually approaching atoms implies an attractive force between them, which is opposed by the increasing electrostatic repulsion of the electrons and nuclei concerned. Overlapping and the consequent output of energy must therefore reach their maxima when a balance is attained between these forces, and the normal interatomic distance in a molecule is so determined. Hence an inverse relation should exist between the length of a bond and its energy of formation, which is illustrated by the data in Table 29.

The spherically symmetrical form of the s orbit is not spatially advantageous to overlapping, unless, as in H_2, a minimum atomic number permits exceptionally close atomic approach. The axial form of p-orbitals, and especially the one-sidedness of hybrid orbitals (fig. 31) is much more favourable to overlap, which is further promoted in hydrides by the small nuclear charge of hydrogen.

Variations among individual bond energies for the three types of orbital conjunction show that bond length cannot solely determine bond energy. At least two other important factors will be: (i) degree of completeness of s-p hybridization, which probably diminishes in the order C, N, O, F; and (ii) change in resonance energy due to the intervention of ionic forms (p. 119). The joint influence of these factors is well illustrated by the series of hydrides in II, and by the series of carbon bonds in III. Each of the atoms in the fluorine molecule contains 3 pairs of spin-coupled, non-bonding electrons in the valency shell, and the mutual repulsion of these pairs has been invoked to explain the exceptional weakness of the F—F bond, but it is still doubtful whether this effect could attain adequate magnitude.

The notion of the overlapping of atomic orbits as the key to chemical union remains incomplete unless we proceed to inquire in more detail into the nature of the *molecular* orbits presumably created by the overlapping. We need to follow the lead given by the successful study of atomic structure, and regard the problem before us as twofold: (*a*) to decide what molecular orbits exist, and how they are related to the atomic orbits, (*b*) to arrange the

Table 29. *Energies of single bonds*

Bond	Bond length (A.)	Bond energy (Cal.)
I. *s—s* bonds		
H—H (H_2)	0·74	103
Li—Li (Li_2)	2·67	26
Na—Na (Na_2)	3·08	17
II. *s—p* (or *s*, *p* hybrid)		
C—H (paraffins)	1·093	98·5
N—H (NH_3)	1·016	93
O—H (H_2O)	0·955	110
F—H (HF)	0·817*	135
III. *p* (or hybrid)—*p* (or hybrid)		
C—C (paraffins)	1·54	83
C—N (alkylamines)	1·47	73
C—O (ethers)	1·42	85·5
C—F (CH_3F)	1·38	95
N—F (NF_3)	1·37	65
O—F (F_2O)	1·38	44
F—F (F_2)	1·44	38†

* Kuipers, Smith and Nielsen, *J. Chem. Physics*, **25**, 275 (1956).
† Sharpe, *Quart. Rev.* xi, 1957, p. 49.

molecular orbits in energy-levels, and so assess and compare their various bonding powers. In solving the formidable but fundamental questions of molecular levels, recourse is had to the idea that the molecular levels will be intermediate between the separate atomic levels and the levels assigned to the total number of electrons in the 'united atom', i.e. a real atom synthesized by the imaginary process of superposing the two nuclei.

Unlike an isolated atom a (diatomic) molecule has an obvious natural axis of reference in the internuclear line. Looking along this direction, *p*-orbits (absolute angular momentum 1 unit), with their axes mutually at right angles, will appear to possess angular momentum λ of the three values ± 1, 0 and 0, according as the axis of rotation lies parallel to the internuclear line or at right angles to it. Molecular orbits may thus be distinguished as π_+, π_-, or σ according as they have respectively apparent values ± 1 or 0 of orbital momentum around the internuclear axis. Since these molecular orbits can by this means be experimentally distinguished, Pauli's principle (pp. 108, 147) allows each to contain two electrons with anti-parallel spins.* In classifying

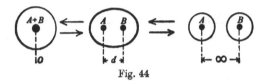

Fig. 44

	United	Bonded	Separate	
	A	F$_2$	F	F
Z	18	18	9	9
d	0	1·44	∞	
	Ne	HF	F	H
Z	10	10	9	1
d	0	0·817	∞	
	P	NO	N	O
Z	15	15	7	8
d	0	1·150	∞	

molecular orbits, λ replaces the quantum number l used for atomic orbits: σ orbits have $\lambda = 0$, and π orbits $\lambda = \pm 1$ (cf. s and p atomic orbits, p. 146).

The correlation of levels in the united atom and the separated atoms is based upon quantum mechanical principles that cannot be entered upon in detail here. A condensed account will be found in Coulson, 'Valence', Oxford 1952 and a detailed exposition in Mulliken, *Rev. Mod. Phys.* 4, 1 (1932). In regard to symmetry *atomic* wave-functions are of two types: they may either change sign when reflected through the nucleus, which lies at their centres of symmetry (type u), or suffer no such change

* The closed *atomic* group p^6 corresponds with the closed *molecular* group $\sigma^2\pi^4$.

(type g). Thus p functions which change sign across their nodal planes, are type u, while s and d functions are of type g. It is a principle of the correlation between the molecule A-A and the united atom that the orbital symmetry be preserved, whence additional information may be gained about the symmetry of molecular orbits in the molecule A_2, which have their centres of symmetry midway between the nuclei. From a physical stand-point the characteristic splitting of levels seen on each side of fig. 45 may be viewed as an internal *Stark effect*, caused by the large *axially* symmetrical electric field developed between the nuclei as they separate from the united atom, or mutually approach in molecule formation. A Stark effect and the related *Zeeman effect* are observed in the spectra of atomic systems when they are caused to radiate in strong applied electric or magnetic fields respectively. Both effects consist of a splitting into several components of the normal spectral frequencies, and therefore of the levels between which electronic transitions occur: and both are caused by the superposition of a field of lower symmetry upon the natural centro-symmetric field of the atom.

The plan below (fig. 45), which is essentially for the combination of two like atoms $A + A$, shows on the right the twice repeated levels in the separated atoms ($d = \infty$), and on the left the levels assigned in the united atom. Owing to the greater nuclear charge in the latter, corresponding levels are deeper in this atom than in the separated atoms. We may regard the levels at an inter-mediate separation d (equal to the actual internuclear distance in the bonded atoms or molecule) as arising either from a slight separation of the nuclei in the united atom, or from the close approach of the separated atoms. Taking the latter standpoint we see that each s-level splits into *two* σ-levels, and each p-level into *four* levels, comprising a pair of σ-levels and a pair of π-levels. As the orbits of the united atom can, in obedience to the Pauli principle, contain only about half of the number of electrons contained in *corresponding* pairs of orbits in the *separated* atoms, about one-half of the electrons must be promoted to levels of higher quantum number when the atoms unite their nuclei ($d = 0$). Since levels in the united atom must lie deeper than corresponding levels in the separated atoms, such promotion may be purely formal, and lead in some cases actually to a *decrease* of energy (e.g. $2p_{\text{atomic}} \to 3s_{\text{united atom}}$ in fig. 45). *Bonding* orbits are identified with *falling* correlation lines, and *anti-bonding* orbits with *rising* lines; it will therefore be noted that a

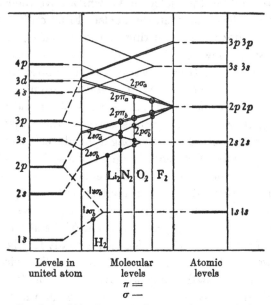

Fig. 45. Correlation of levels in united atom and separated like atoms.
(After Herzberg, *Z. Physik*, **57**, 601 (1929), and
Mulliken, *Rev. Mod. Phys.* **4**, 1 (1932).)

'promoted' molecular orbit is not *necessarily* anti-bonding. The *absolute* level of an anti-bonding orbit (e.g. $2s\sigma_a$) may lie below those of bonding orbits (e.g. $2p\pi_b$). Fig. 46 shows that only minor changes in correlation are needed for two atoms A and B with comparable nuclear charges (N and O, or C and O). We shall use the suffixes a and b to classify orbits into anti-bonding and bonding respectively, as has been done in the plan above.

When the atomic orbits are full to capacity, i.e. are closed shells, equal numbers of electrons pass into bonding and anti-bonding orbits respectively on approach of two atoms A, and no chemical union can be achieved. We see at once why the molecules

He$_2$ and Be$_2$ cannot normally exist;* each would contain one filled bonding orbit ($1s\sigma_b$ for He and $2s\sigma_b$ for Be) and one filled anti-bonding orbit ($1s\sigma_a$ and $2s\sigma_a$ respectively), the latter being the stronger. On the other hand, both electrons in H + H, if with opposed spins, pass into the bonding orbit $1s\sigma_b$, and a stable molecule H$_2$ results. Further, the molecular ions He$_2^+$ and HeH$^+$

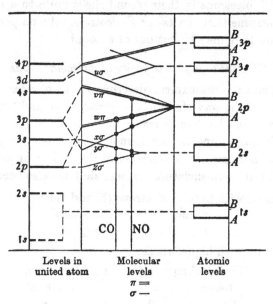

Fig. 46. Correlation of levels in united atom and
separated unlike atoms (A, B).

can have existence, since the former will contain two electrons in the bonding orbit $1s\sigma_b$ and only one in the anti-bonding orbit $1s\sigma_a$; HeH$^+$ has the same configuration as H$_2$ ($1s\sigma_b^2$). The reactions leading to the formation of these molecular ions are probably

   (1)   He $(1s.2s)$ + He $(1s^2)$ = He$_2^+$ $(1s\sigma_b^2.1s\sigma_a)$ + ϵ,

   (2)   H$_2^+$ + He = HeH$^+$ + H.

This viewpoint should be compared with the earlier, more qualitative, treatment by localized pairs on p. 112. As is indicated

* An excited He$_2$ molecule is recognizable spectroscopically (Curtis, *Trans. Faraday Soc.* 25, 694 (1929), and Arnot and M'Ewen, *Proc. Roy. Soc.* A, 171, 106 (1939).)

in the plan the splitting of $2p$ atomic orbits occurs at larger internuclear distances than the splitting of $2s$-orbits, and $1s$-orbits remain unmodified up to quite small distances. In terms of our previous discussion (p. 179), we may say that the order of overlapping with decreasing distance is $2p$, $2s$, $1s$. Hence in general p-orbits play the most important part in bonding. When atoms with electrons in their second shells unite to a molecule we may assume that the $1s$- or K-electrons take no part in the bonds, and they may be left out of account.

Classification and capacity of molecular orbits

The quantum mechanical descriptions with regard to the symmetry of the wave-functions—'gerade' $= g$, and 'ungerade' $= u$—applicable to homonuclear molecules are given. When the combining atoms are not homonuclear it is appropriate to employ a more general description of the molecular orbits; for this Mulliken's nomenclature ($z\sigma$, $w\pi$, etc.) has also been added.

(1) Atoms with only $1s$-electrons (H and He):

Bonding	Anti-bonding
$1s\sigma_b$ 2	$1s\sigma_a$ 2

(2) Atoms with higher shells:

Non-bonding $1s + 1s$ 4 (K-electrons)

Bonding				Anti-bonding			
$2s\sigma_b$,	$2s\sigma_g$,	$z\sigma$	2	$2s\sigma_a$,	$2s\sigma_u$,	$y\sigma$	2
$\left.\begin{matrix}2p\pi_{b+}\\2p\pi_{b-}\end{matrix}\right\}$,	$2p\pi_u$,	$w\pi$	$\begin{cases}2\\2\end{cases}$	$\left.\begin{matrix}2p\pi_{a+}\\2p\pi_{a-}\end{matrix}\right\}$,	$2p\pi_g$,	$v\pi$	$\begin{cases}2\\2\end{cases}$
$2p\sigma_b$,	$2p\sigma_g$,	$x\sigma$	2	$2p\sigma_a$,	$2p\sigma_u$,	$u\sigma$	2

It might plausibly be objected that atomic electrons would not spontaneously adopt orbits of the anti-bonding type, but would remain in atomic orbits at a lower energy-level. There seems in fact to be too sharp a contrast between Ne + Ne, where all the electrons remain in atomic orbits, and F + F, where our theory seems to allocate at least all the ten $2p$-electrons to molecular orbits. It should, however, be remembered that if bonding actually occurs the permanent contiguity of the atoms must profoundly modify most of the originally atomic orbits. In any case all electrons retaining atomic orbits in a molecular structure would exert mutual repulsion, i.e. an anti-bonding effect. In support of the conception of anti-bonding molecular orbits as a working hypothesis it may be urged that they very greatly simplify the interpretation of the excited states of molecules, in which electrons have been transferred from lower to higher energy states.

On ionization a molecule loses electrons first from the highest molecular levels, and then from other levels in order of depth. When anti-bonding orbits lie highest electrons will generally be first removed from these, and the molecular ion (e.g. O_2^+) should then be more stable than the neutral molecule: the increased net effect of the bonding electrons should also lead to a decreased internuclear distance. Table 30 shows that both these predictions from the theory of molecular orbits are well substantiated. The data refer to ground-states of molecules and ions. (D = energy of formation, Cal.)

Table 30

Molecule	Observed molecular state	D (Cal.)	d (A.)	Electron ionized	Net bonding electrons
H_2	$^1\Sigma_g^+$	103	0·74	$1s\sigma_b$	2
H_2^+	$^2\Sigma_g^+$	62	1·06		1
O_2	$^3\Sigma_g^-$	117	1·207	$2p\pi_a$	4
O_2^+	$^2\Pi_g$	150	1·123		5
NO	$^2\Pi$	150	1·15	$v\pi$	5
$(NO)^+$	—	—	1·07	—	6
Cl_2	$^1\Sigma_g^+$	58	1·989	$3p\pi_a$	2
Cl_2^+	$^2\Pi$	(100)	1·89	—	3

The co-existence of anti-bonding with bonding orbits is a consequence of the application of wave-mechanical theory. If Ψ is a molecular wave-function of a homonuclear molecule AB, occupied by two electrons, then at any point in the molecular space Ψ^2 defines precisely the density of electronic charge at that point. Suppose the electrons, or alternatively the nuclei to be interchanged in position; because of its symmetry the molecular constitution remains unaltered, and Ψ^2 must therefore also be unchanged. Since Ψ^2 arises either from $\Psi \times \Psi$ or $(-\Psi) \times (-\Psi)$, we deduce that all acceptable molecular wave-functions must behave either *symmetrically* to such interchange, that is remain unaltered, or *antisymmetrically*, that it become multiplied by -1. No other changes can be permitted.

If ψ_A and ψ_B are *atomic* wave-functions, the electronic arrangements A(1).B(2) and A(2).B(1) are respectively represented by the product functions $\psi_A(1).\psi_B(2)$ and $\psi_A(2).\psi_B(1)$, where the numerals denote the two members of the electron pair. Each product function has the same probability, but they are not acceptable wave functions of AB since neither is symmetrical or antisymmetrical for particle interchange. To introduce the necessary property in Ψ we form the combinations

$$\Psi_b = c[\psi_A(1)\,\psi_A(2) + \psi_A(2)\,\psi_B(1)],$$

$$\Psi_a = c[\psi_A(1)\,\psi_B(2) - \psi_A(2)\,\psi_B(1)],$$

where c is a numerical (normalizing) constant. These forms satisfy not only the symmetry property but also the intuitive demand that the product functions of equal probability be given parity of treatment. When explicit forms are given to ψ_A and ψ_B and Ψ_a and Ψ_b calculated, it is found that Ψ_b remains large between the nuclei of A and B, while Ψ_a sinks in this region to zero. Thus Ψ_b tends to draw A and B together into a molecule, and Ψ_a to separate the molecule into atoms.

Having gained a means of arranging molecular orbits in their order of energy we seek to complete the analogy with atomic systems by attempting a geometrical representation of the molecular types. While s wave-functions have everywhere the same sign ($+$ or $-$), p (and d) functions change sign across their nodal planes, which include the nucleus. Bonding molecular orbits result when the contiguous parts of the two atomic orbits have like signs, and the anti-bonding type when the signs are opposed. We see here a representational form of the distinction between the two types of molecular orbit. By introducing appropriate pairs of atomic functions into the expressions given above, and then calculating the contours of Ψ_a and Ψ_b, it is found that everywhere in the internuclear space Ψ_b^2 exceeds, and Ψ_a^2 is less than, the sum $\psi_A^2 + \psi_B^2$, which measures the electron charge density resulting from superposing the atomic orbits. Since the total integrated density is constant and must amount to just two electron charges, it is clear that in the formation of molecular orbits electron density drifts inwards or outwards according as the orbit is bonding or anti-bonding.

Two p orbits approaching along their common axis ('head on') give rise to $p\sigma$ orbits which clearly acquire axial symmetry round the internuclear line. As the internuclear distance decreases, and because of the electronic drift mentioned above, the 'off-side' lobes in the representation of the p orbits become steadily diminished in a $p\sigma_b$ orbit (see, for example, fig. 51, p. 195) and enhanced when the orbit is anti-bonding ($p\sigma_a$). These changes would reach their limit in the united atom when $p\sigma_b$ and $p\sigma_a$ become respectively s and p atomic orbits (see fig. 45). In heteronuclear diatomic molecules such as the hydrogen halides, an orbit similar to $p\sigma_b$ arises from the conjunction of a p orbit in the halogen with an s orbit in the hydrogen. Such a molecular orbit can have only one 'off-side' lobe (see fig. 30) and on this account is unsymmetrical about the median plane. We can see that such unsymmetry creates a dipole in the direction $H_+ - Cl_-$, and we can predict a certain polarity of the molecule in the expected direction, even if explicit ionic terms are omitted from the molecular function (see p. 119). What proportion of the whole observed magnitude of the dipole strength can be thus explained is a matter of detailed calculation, which at present awaits conclusion.

Molecular orbits $p\pi_{a,b}$ are formed by the approach of p orbits with their axes *parallel* ('side-ways on', fig. 53, p. 196) and, for a given internuclear distance, this conjunction is less favourable to overlapping than the 'head on' approach, although each lobe plays an equal part. For this reason the degenerate $p\pi$ orbits lie between the $p\sigma$ orbits on the energy scale (see fig. 45). Bonding $p\pi$ orbits have only one nodal plane, which passes through the internuclear line, and naturally become p orbits in the united atom, while $p\pi_a$ orbits, with a second nodal plane orthogonal to the first, become d orbits at this limit. From the orbital constitution of the N_2 molecule as portrayed in fig. 45 we can infer

a general representation of multiple bonds, of which it is clear
that the member links are not all equivalent. The strongest link,
of type $p\sigma$, might be regarded as formed from $[sp]$ hybrid orbits
without essentially changing the picture, but the remaining $p\pi_b$
orbits can be produced only if '*pure*' p orbits are available in the
atoms. Although this scheme is fundamental to an understanding of multiple links it cannot be directly correlated with

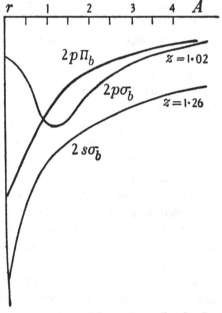

Fig. 47. (Reproduced from Lennard-Jones, *Trans. Faraday Soc.* **25**, 678 (1929).

known bond energies, for the observable links $\rangle N - N \langle$ (1·47 A.)
and —N═N— (1·20 A.) obviously differ from the member links
in N≡N (1·09 A.), especially in length.

Figs. 45 and 46 indicate the correlation between the lower levels
occupied by electrons in a pair of light atoms A and B of equal
(or nearly equal) atomic number Z, and the levels they must
assume when the nuclei of A and B coalesce to yield the 'united
atom'. The scheme shows in a general way which electrons will
exert an anti-bonding effect in the molecule AB, and which, by
adopting deeper orbits, will have a bonding effect. To gain even

an approximate knowledge of the *actual* energy of the molecular levels in a particular case requires a much more detailed consideration than figs. 45 and 46 might suggest. Fig. 47 shows graphically how bonding levels arising from two lithium atoms ($Z = 3$), and leading to the formation of the molecule Li_2, actually change their positions and mutual relations as the internuclear distance r is progressively reduced. In performing the necessary calculations* it was assumed that the innermost $1s$-electrons played

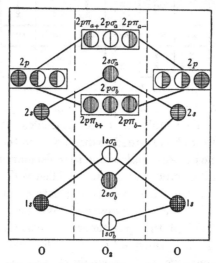

Fig. 48. Molecular orbits in the oxygen molecule O_2.

no role beyond screening the nuclear charge from the active $2s$-electrons in the second atomic shells. The screening effect, expressed as the effective nuclear charge z shown in the figure, varies with the type of orbit occupied by the bonding electrons.

Another mode of exhibiting the relations of atomic to molecular orbits is illustrated by fig. 48, which indicates the electron assignments in the molecule O_2 and how they originate in the two oxygen atoms. Molecular orbits must be occupied by electrons drawn only from the *appropriate* atomic orbits indicated in the scheme. Hydrogen is probably the only common stable molecule in which electrons in $1s$ atomic orbits are transferred to molecular orbits

* Lennard-Jones, *Trans. Faraday Soc.* **25**, 678 (1929).

($1s\sigma_b$). Hence in other molecules the molecular orbits $1s\sigma_b$ and $1s\sigma_a$ must be left untenanted, as shown in fig. 48.

Table 31. *Distribution of electrons in molecular orbits*

	Bonding orbits				Anti-bonding orbits			Ground state	$\dfrac{n_b - n_a}{2}$
	$2s\sigma_b$	$2p\sigma_b$	$2p\pi_{b+}$	$2p\pi_{b-}$	$2s\sigma_a$	$2p\pi_{a+}$	$2p\pi_{a-}$		
C$_2$	2	1	2	1	2	—	—	$^3\Pi$	2
N$_2$	2	2	2	2	2	—	—	$^1\Sigma$	3
O$_2$	2	2	2	2	2	1	1	$^3\Sigma$	2
F$_2$	2	2	2	2	2	2	2	$^1\Sigma$	1
	$z\sigma$	$x\sigma$	$w\pi_+$	$w\pi_-$	$y\sigma$	$v\pi_+$	$v\pi_-$		
CO	2	2	2	2	2	—	—	$^1\Sigma$	3
NO	2	2	2	2	2	1	—	$^2\Pi$	$2\frac{1}{2}$

(Mulliken, *Rev. Mod. Phys.* 4, (1932).)

The scheme above shows the distribution of the disposable electrons in some simple diatomic molecules. The $1s$- or K-electrons are not to be regarded as disposable in these formations (see above). The process of allocation to molecular orbits is almost mechanical in its simplicity; orbits are filled with their complements of two electrons from the appropriate atomic orbits until all available electrons are exhausted. We may contrast this method with that of the spin theory by considering the case of F$_2$. Basing our principles on a qualitative relation with the inert elements we confer a single electron-pair bond on F$_2$ because each atom contains one unrestricted electron (see p. 110), and we then ascribe the whole bonding effect to this electron-pair. In the other method we assume that there are actually four fully occupied bonding orbits whose bonding power is offset by three fully occupied anti-bonding orbits, giving one net bonding orbit. It is probable that the actual condition of (diatomic) molecules is one intermediate between the two extreme representations. For molecules containing multiple bonds, in which the internuclear distance d_e is less than $2d_0$, where d_0 is the 'covalent radius' (p. 45), the model suggested by the method of molecular orbits is probably near the truth, while for molecules with longer internuclear distances ($d_e \sim 2d_0$) the method of localized pairs gives the truer picture. Thus for C$_2$, N$_2$ and O$_2$, with $d_e/d_0 = 1\cdot70$, $1\cdot55$ and

1·60 respectively, we use the method of molecular orbits: for F_2, $(d_e/d_0 > 2)$, the method of localized pairs is a good approximation. Herein we have the reason why the (single bond) covalent radii of elements are so much more nearly constant than the lesser radii of multiple bonding (p. 45). In diatomic hydrides (e.g. HF) the molecular levels lie close to those of the united atom.

The example of the oxygen molecule (fig. 48) manifests in a striking way the superior powers of the method of molecular orbits. In assigning the last two electrons to the $2p\pi_a$-orbits we may (1) put both in one orbit, when the spins must be coupled to zero, (2) place one in each orbit, with or without coupling between the two spin momenta. The correct assignment will be that of least energy. It is actually found on a detailed consideration* that the arrangement (2) with *orbital* momenta coupled to zero and *spins* parallel (total spin momentum 1 unit, $^3\Sigma$) lies about 25 Cal. lower than the alternative choice ($^1\Delta$). The resulting molecule has resultant spin momentum of 1 unit in the ground-state and is therefore *paramagnetic* in this state. The electron-pair method, in demanding the coupling of the spins of all electrons in the molecule, fails to reproduce the most outstanding physical property of the O_2 molecule.† On tracing back the electron spins shown by fig. 48 for molecular oxygen to the constituent atoms, these are seen not to have identical electronic states: in one the spins of the 'lone' electrons are parallel ($^3P_0$) and, in the other, anti-parallel ($^3P_1$). Both these states belong, however, to the ^3P ground state of atomic oxygen (see p. 151). The case of NO, in which only one $2p\pi_a$-orbit can be (singly) occupied, has already been mentioned (p. 159). The energy difference (120 cm.$^{-1}$) between the states $^2\Pi_{\frac{1}{2}}$ and $^2\Pi_{\frac{3}{2}}$ being only $0·6kT_{300°}$, the molecular magnetic moment μ (p. 160) varies with T, as found experimentally.‡ The diamagnetic state $^2\Pi_{\frac{1}{2}}$ lies lowest, and is stable at low temperature. The similarity of the electronic distribution

* Lennard-Jones, *Trans. Faraday Soc.* 25, 668 (1929).

† The molecule S_2 is also paramagnetic (Shaw and Phipps, *Phys. Rev.* 38, 174 (1931)).

‡ van Vleck, *Phys. Rev.* 31, 517 (1928); Stössel, *Ann. Physik*, 10, 393 (1931).

in CO and N_2 agrees with the less direct deduction from the spin theory (p. 123). Symbols indicating the states of molecules are closely similar to the convention for atomic states (p. 149). Thus a Σ state is analogous to an S atomic state in having no resultant angular momentum λ round the internuclear line: a Π state has one such unit, analogous to a P state for atoms, and so on. The molecular multiplicity is again placed at the upper left of the symbol, and records the resultant spin momentum S as $2S+1$.

The development and present relations of the three main methods of attacking the problem of chemical union may be thus summarized. Heitler and London's original method applied to combination involving only s atomic orbits, that is, effectively only to H_2, for which accurate results are obtained but only by highly refined calculation*. The spin theory or localized pair treatment (Pauling-Slater), rendered sufficiently flexible by the essential device of 'resonance', can offer a very broad survey of molecular structure along familiar chemical lines, but cannot render quantitative details that may be of crucial importance (cf. O_2 above). The method of molecular orbits, originally confined to diatomic molecules has developed into a powerful instrument applicable to complex molecular systems.

As a penetrating comment on the techniques of attacking problems of molecular structure we may quote from Lennard-Jones: '...the electrons constituting the bonds cannot be segregated into closed localized pairs. This feature is represented in the Pauling method by the superposition of a number of canonical structures, each of which corresponds to a chemical picture of localized bonds. The state of the molecule has properties which are different from those of the individual canonical structures, but can be defined or interpreted in terms of a set of them. The energy of the lowest or normal state is usually lower than that of any one canonical structure, even than that which would appear from the orthodox method of drawing bonds to be the most stable.'†

* James and Coolidge, *J. Chem. Physics*, **1**, 825 (1933).
† *Proc. Roy. Soc.* A, **158**, 280 (1937).

The following general accounts related to the theory of molecular orbits provide a conspectus of its development:

Lennard-Jones and Garner, *Trans. Faraday Soc.* **25**, 611 (1929).
Lennard-Jones, *ibid.* p. 668.
Herzberg, *Z. Physik*, **57**, 601 (1929).
Mulliken, *Rev. Mod. Phys.* **4**, 1 (1932).
Lennard-Jones, *Trans. Faraday Soc.* **30**, 70 (1934).
van Vleck and Sherman, *Rev. Mod. Phys.* **7**, 167 (1935).
Herzberg, *Molecular Spectra and Molecular Structure:* (1) *Diatomic molecules* (Prentice and Hall, 1939), Chap. VI.

Single and multiple bonds involving carbon——Unsaturation

If one CH_3 group in $CH_3.CH_3$ is imagined to be rotated round the C—C axis while the second remains at rest the energy of the molecule undergoes regular fluctuation (fig. 49a):

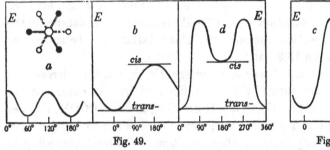
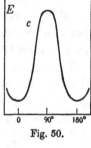

Fig. 49.   Fig. 50.

The maxima probably occur when the six H atoms form a regular trigonal prism, and the minima when they form a trigonal anti-prism, i.e. are 'staggered' as shown (fig. 49a). The energy-difference between the maxima and minima is small (about 2 Cal.), but is manifested in the change of specific heat with temperature.* When the same procedure is applied to 1 : 2 dichlorethane (ethylene dichloride) the energy changes are larger† (fig. 49b) and the *trans-* position is definitely preferred, as is shown by the difference between the electric moment calculated

* Stitt, *J. Chem. Physics*, **7**, 297 (1939); Kistiakowsky, Lacher and Stitt, *J. Chem. Physics*, **7**, 289 (1939).

† Wu, *J. Chem. Physics*, **7**, 965 (1939) (for a critical discussion of rotation in ethane and halogeno-ethanes see Glockler, *Rev. Mod. Phys.* **15**, 'The Raman Effect', Section IV, p. 145 (1943)).

for free rotation (2·54) and the actual moment (1·12 at 305° K.).*
Evidently the interaction of the two Cl atoms is the main cause
of a certain resistance to turning. For ethylene the energy-
rotation relation is shown in fig. 50. The energy-difference
has risen to about 20 Cal. (see below) and the derivative
CHD=CHD must exist in *cis*- and *trans*- forms, of equal stability.
Here the strong resistance to rotation must be wholly ascribed
to the nature of the C, C linkage. In CHCl=CHCl (fig. 49*d*)
and COOH.CH=CH.COOH we have both causes of resistance
to rotation superimposed and the *cis*- and *trans*- forms have
different stability. The energy required to rotate one =CH.COOH
group through 180° in maleic or fumaric acid is 15·8 Cal.†
(the heat of activation of the *cis*-, *trans*- change), while the
energy-difference between the two forms is 6–7 Cal. (from the
heats of combustion of the two acids). Since in maleic acid the
repulsive interaction of the COOH groups offsets resistance of the
bond to rotation, the energy of this rotation must be somewhat
greater than 15·8 Cal.‡

The heat ΔH of *cis-trans* isomerization in ethylene derivatives
C*ab*=C*ab* is naturally much influenced by the nature of the
groups *a* and *b*. When *a* = H or alkyl, and *b* = COOH, ΔH lies
between 6 and 10 Cal./mol.; but when *a* = H and *b* = Cl or Br,
ΔH is usually much smaller (less than 1 Cal./mol. according to
Maroney, *J. Amer. Chem. Soc.* **57**, 2397 (1935), Wood and
Stevenson, *ibid.* **63**, 1650 (1941) and Noyes and Dickinson, *ibid.*
65, 1427 (1943)).

The carbonylic, ethylenic and acetylenic bonds

When in each of the two carbon atoms one of the 2*s*-electrons
has been promoted to a 2*p*-orbit and the quadrivalent condition

* Zahn (*Phys. Rev.* **40**, 291 (1932)) gives the temperature variation
of the moment as 1·12$_{305°}$ to 1·54$_{554°}$.

† Højendahl, *J. Physical Chem.* **28**, 758 (1924).

‡ It is of interest to note that an X-ray examination of the solid ester,
dimethyl fumarate

CH$_3$.OOC.CH
‖
HC.COOCH$_3$

has directly proved that this *trans*- isomer is 'nearly, if not quite,
planar' (Knaggs and Lonsdale, *J. Chem. Soc.* 1942, p. 417).

is thereby established (p. 114), the most natural way in which to envisage the formation of a double bond between them is illustrated in fig. 51, in which some attempt has been made to indicate that the four C,H bonds result from overlapping of p_C- with s_H-orbits, while the double linkage is formed by the engagement of p_C with p_C and s_C with s_C. The planes CH_2 would be parallel and at right angles to the C—C internuclear line. This model, however, cannot be that of ethylene, for since the s-orbits are spherically symmetrical the energy of the link

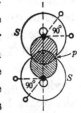

Fig. 51.
●, nucleus of C;
o, nucleus of H.

does not appreciably depend on the relative orientation of the CH_2 groups, i.e. 'free' rotation occurs. When we proceed not only to promote an s-electron in each C atom but to hybridize the sp^3 group to $[sp^3]$ so that all four orbits become equivalent (p. 115), then a model as in fig. 52 suggests itself. In this model both members of the double bond are equivalent, and must be associated with equal bonding energies. As in the classical tetrahedral model all nuclei are held co-planar. This model, however, goes to the opposite extreme of rigidity, for the energy required to rotate from the *cis-* to the *trans-* position would be much greater than is actually found in such transformations experimentally.

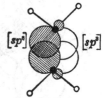

$[sp^3]$ $[sp^3]$

Fig. 52.
●, nucleus of C;
o, nucleus of H.

It has already been mentioned (p. 115) that there are two modes of hybridizing sp^3 to give configurations of almost equal energy: (a) the *tetrahedral* mode, used in fig. 52, (b) the *trigonal* mode. In the latter the hybridization is practically confined to the group sp^2, and a pure p-orbit occurs directed at right angles to the plane in which the axes of the three hybrids $[sp^2]$ lie. It must be emphasized that this mode of hybridization is inseparable from the formation of a 'double' (i.e. ethylenic) bond: in fact the hybridizing process is part of the construction of the ethylenic model. In this, one of the three hybrid orbits binds the carbon atoms strongly together and the other two bind hydrogen on each carbon atom. Rotation round the C—C axis is resisted solely by the weak overlap of the pure p-orbits directed at right angles to the H_2C—CH_2 plane (fig. 53a). It may be estimated that the energy (of activation) required for a *cis-trans* conversion in CHD=CHD bonded in this way is only about 20 Cal.* In the tetrahedral model (fig. 52) we have two moderately strong bonds of equal strength, while in the last model we have one very strong bond (σ) and one weak bond (π); the total bond energies decisively favour the last model.† Immediately addition occurs to the ethylenic linkage *tetrahedral* hybridization automatically sets in (fig. 53b). As the model (a, fig. 53) contains no uncoupled electron spins, ethylene is diamagnetic, in agreement with experiment.

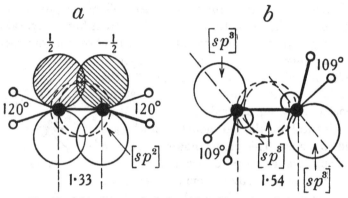

Fig. 53. Orbits in normal ethylene (a), and in active ethylene (b), showing normal, i.e. *trans*- addition.

The characteristics of the tetrahedral and trigonal types of bond are well contrasted in diamond (p. 40) and graphite. The latter consists of macromolecular sheets of co-planar carbon atoms bonded by 'trigonal' bonds within the sheet, but not bonded chemically between the sheets. The lattice of a sheet presents the hexagonal ('honey-comb') appearance of fig. 54. Thus in place of the great isotropic hardness of diamond we have in graphite great hardness only in a direction tending to compress the sheets, which easily slide over each other. The lubricating properties of

* Penney, *Proc. Phys. Soc.* 46, 333 (1934).
† Penney, *Proc. Roy. Soc.* A, 144, 166 (1934).

graphite are thus explained. Further, it is no longer permissible to consider the 'vertical' p-bonds to be discretely paired as in ethylene, but it must be assumed that they are held in common by all the atoms in the sheet. In this we confer upon graphite in respect to its vertical p-bonds the character of a metal, and the electrons in these p-bonds are conduction electrons. Moreover, the mobility of the electrons will lead to optical opacity, also a metallic character, and opposed to the transparency of diamond.

We may extend these conceptions to explain the properties of benzene, which may be treated structurally as an isolated hexagonal unit of the graphite sheet. In the hydrocarbon six

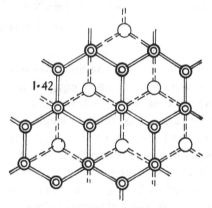

Fig. 54. The structure of graphite.

'mobile' electrons will be present, and here again it is not permissible to pair them off into three discrete ethylene linkages. As the mobile electrons are now confined permanently to one molecule, electrical conduction is not possible, but the correctness of the structural picture is strongly supported by the remarkable anisotropy of benzene (and other benzenoid hydrocarbons) to magnetic fields.* Further, we expect, as is actually found, that all the C, C distances in benzene are equal and the molecule has the form of a plane regular hexagon.

* Krishnan and co-workers, *Phil. Trans.* **231**, 235 (1933); *ibid.* **234**, 265 (1935); Krishnan and Ganguli, *Z. Krist.* **100**, 530 (1939): summarized by Stoner, *Magnetism* (Methuen, 1936), p. 40.

We may now proceed still further along these lines by considering the hydrocarbons hexatriene $CH_2:CH.CH:CH.CH:CH_2$,
butadiene $CH_2:CH.CH:CH_2$, and finally ethylene $CH_2:CH_2$.
Detailed calculation* predicts a progressive discrimination in
bond-lengths:

Bond-lengths

	n	'Single'	'Double'	
Ethylene	2	—	1·33	(experimental value)
Butadiene	4	1·41	1·34	
Hexatriene	6	1·40	1·35	calculated
	∞	1·38	1·38	
Benzene	—		1·39	(See footnote)†
Graphite	—		1·42	(from X-ray analysis) Lipson and and Stokes, *Proc. Roy. Soc.* A, **181**, 101 (1942)

The closed nature of the benzene ring gives to its distances the
equality only reached in an open (conjugated) chain of infinite
length. As the number n of carbon atoms in the so-called
conjugated chain diminishes, definite single and double bonds
tend to appear, thus implying a tendency to discrete pairing of
the vertical p-orbits as in ethylene itself. Calculations based on
the assumption of one 'mobile' electron per C atom predict
significantly different C, C lengths in graphite and benzene, in
excellent agreement with the observed values.‡

It has long been recognized that a conjugated system of bonds
(alternating ethylenic and single bonds) shows specific properties,
indicating that the ethylenic bonds are not independent. Thus in
the addition of chlorine to butadiene the two addition products
$ClCH_2—CH:CH—CH_2Cl$ and $ClCH_2—CHCl—CH:CH_2$ are
produced simultaneously. This type of 1:4 addition continues
to be shown when one of the members of the conjugated system
is $C:O$. Thus hydroxylamine yields with the terpene pulegone (I)

* Lennard-Jones, *Proc. Roy. Soc.* A, **158**, 280 (1937).

† (ED, vapour) Karle, *J. Chem. Physics*, **20**, 65 (1952), (X-ray, solid)
Cox & Smith, *Nature*, **173**, 75 (1954).

‡ Coulson, *Nature*, **154**, 797 (1944); *Proc. Roy. Soc.* A, **169**, 413
(1938).

the hydroxylamino-oxime (II):

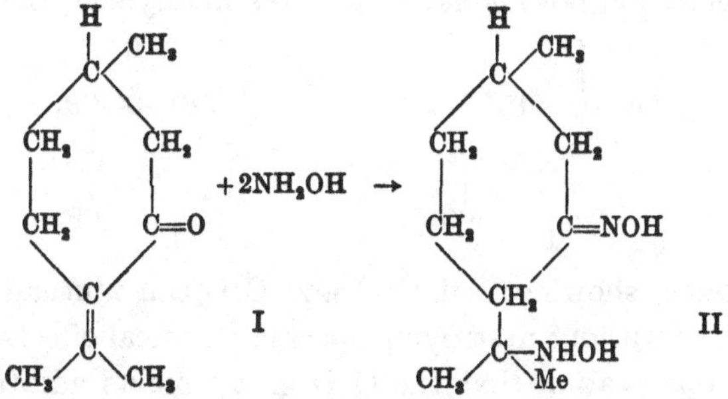

$+ 2NH_2OH \rightarrow$

I   II

whose formation is comprehensible if one molecule of NH_2OH undergoes 1:4 addition:

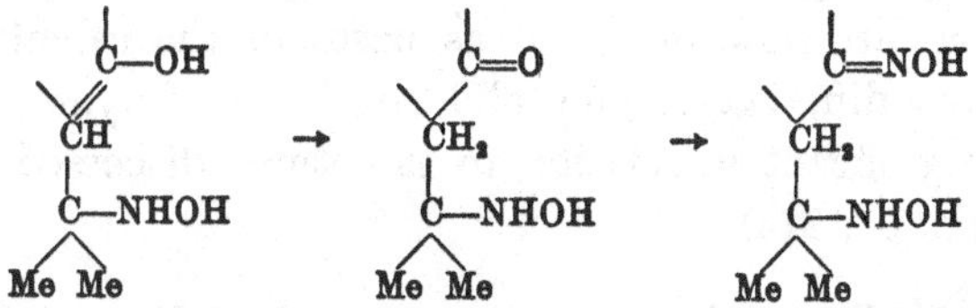

The expectation that oxalic acid and its ion would show effects of conjugation in the system $O{=}C{-}C{=}O$ is however disproved by recent structure determinations*

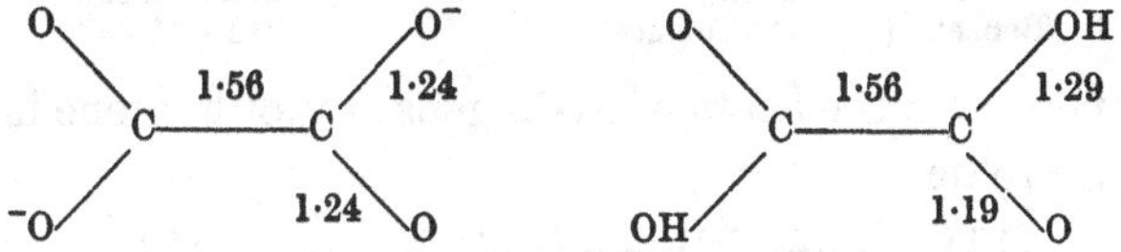

Comparison with the normal lengths $C{=}O$, 1·22A., $C{-}O$, 1·42A. and $C{-}C$, 1·54A. shows that in the ion full electronic mobility occurs within the $(CO_2)^-$ group without extending to the $C{-}C$ link. In the acid also the C, C link is clearly a single bond. It is clear that if the vertical p-orbits in butadiene are to form discrete ethylenic bonds the necessary pairing can occur in two ways, either between the 1:2 and 3:4 carbons, or between the 2:3 carbons, leading to 1:2 (or 3:4) addition or 1:4 addition respectively. Thiele's theory of this behaviour, proposed empirically, is seen to bear much resemblance to

* Cox, Dougill and Jeffery, *J. Chem. Soc.* 1952, pp. 4854 and 4864.

modern formulations. Glutaconic and acetoacetic esters (I and II respectively) both contain 'reactive methylene' CH_2:

$$\text{EtO—C(=O)—CH}_2 \qquad\qquad \text{EtO—C(=O)—CH}_2$$
$$\text{CH=CH} \qquad\qquad\qquad \text{C=O}$$
$$\text{COOEt} \qquad\qquad\qquad \text{CH}_3$$
$$\text{I} \qquad\qquad\qquad\qquad \text{II}$$

These facts, showing that $C:O$ and $C:C$ can mutually replace each other without destroying specific chemical effects, and the further observation that the $C:O$ group reacts additively like $C:C$ (with suitable addenda), strongly suggest that the nature of the bonding is similar in the two groups.

Heats of (catalytic) hydrogenation, being smaller than heats of combustion, are now preferred as bases of the calculation of small energy differences. The following heats of hydrogenation are very significant in relation to structures discussed above* (see also pp. 68, 214):

Table 32. *Heat of hydrogenation per mol. of H_2 (Cal. mol.)*

Ethylene, $CH_2:CH_2$	$32 \cdot 58 \pm 0 \cdot 05$
Butadiene, $CH_2:CH.CH:CH_2$	$28 \cdot 53 \pm 0 \cdot 075$
Pentadiene, $CH_2:CH.CH_2.CH:CH_2$	$30 \cdot 39 \pm 0 \cdot 075$
Dihydrobenzene	$27 \cdot 68 \pm 0 \cdot 05$
Tetrahydrobenzene	$28 \cdot 59 \pm 0 \cdot 10$
Benzene (and homologues)	$16 \cdot 1 - 15 \cdot 9$

From these data the heats of hydrogenation of benzene in stages are found to be

$$C_6H_6 \xrightarrow[+7\cdot4]{} C_6H_8 \xrightarrow[-26\cdot8]{} C_6H_{10} \xrightarrow[-28\cdot6]{} C_6H_{12} \text{ (Cal. per mol.)}$$

The first stage, to dihydrobenzene, is thus *endothermal*. Taylor and Turkevich† have calculated from thermal data the changes in *free* energy $\Delta F°$ for the stages in the reduction of benzene

$$C_6H_4 \xrightarrow[]{+14\cdot9} C_6H_8 \xrightarrow[]{-17\cdot5} C_6H_{10} \xrightarrow[]{-19\cdot2} C_6H_{12} \text{ (cyclohexane)}$$

($\Delta F°$ in Cal.). In relation to these facts it may be observed

* Kistiakowski *et al., J. Amer. Chem. Soc.* 57, 65, 876; 58, 137, 146; 59, 831, and *Ann. Reports*, 1937, p. 214.

† *Trans. Faraday Soc.* 35, 923 (1939).

that by the usual catalytic methods of hydrogenation, however moderated, it has not yet proved possible to obtain from benzene directly the partially hydrogenated products dihydro- and tetrahydrobenzene, although these compounds are readily prepared by well-known methods from hexahydrobenzene (see further, p. 216).

If we take the triple bond in nitrogen as a model for that in acetylene, then in each carbon atom two pure, singly-occupied p orbits must be reserved to form the two π members of the bond, leaving in each atom only an s and one p orbit to be hybridized, to two $[sp]$ orbits. Such hybrids have an unsymmetrical pattern similar to that of $[sp^2]$ and $[sp^3]$, but their axes are directed oppositely along the internuclear line with a mutual angle of 180°. In common with $[sp^2]$ and $[sp^3]$ their superior powers of overlapping in bond formation promote their use in N_2, HCN, C_2H_2, and by the central carbon atoms in CO_2, allene, $CH_2{=}CH{=}CH_2$, and ketene, $CH_2{=}C{=}O$. This model for acetylene suggests that it should display chemical properties similar to those of ethylene. In fact, acetylene is more stable to certain addition reagents than ethylene. The discriminating reaction

$$CH_2{=}CH{-}CH_2C{\equiv}CH + Br_2 \ = \ CH_2Br{-}CHBr{-}CH_2{-}C{\equiv}CH$$

illustrates its relatively slow reaction with halogens: hydrogen halides also react very slowly, and acetylene is more stable than ethylene to oxidants. On the contrary, the reaction

$$CH_2{=}CH{-}CH_2{-}C{\equiv}CH + H_2 \ = \ CH_2{=}CH{-}CH_2{-}CH{=}CH_2,$$

promoted by suitable catalysts, shows that acetylene is more rapidly hydrogenated than ethylene. Lastly, acetylene, by forming acetylides (p. 54) reveals an acidic function quite absent in ethylene.

Such distinguishing chemical features suggest that the C atoms in C_2H_2 are more electronegative than those in C_2H_4, and this supposition is corroborated by less qualitative evidence. In Table 33 are seen the lengths and stretching-force constants* of a series of X—H bonds in which it may safely be assumed that

* Linnett, *Quart. Rev.* 1, 73 (1947).

Table 33

Bond	C—H (CH$_4$)	C—H (C$_2$H$_2$)	N—H (NH$_3$)	O—H (H$_2$O)	F—H (HF)
Force constant (dyne/cm. × 10^{-5})	5·0	5·9	6·5	7·6	9·7
Bond-length (A.)	1·094	1·059	1·014	0·958	0·817

X increases steadily in electronegative character. The steady rise of the force constant and the concomitant decrease of bond length will be noted, and the significant position of C—H in C$_2$H$_2$ between C—H in CH$_4$ and NH$_3$. In the series of methyl halides in Table 34 we see a similar, if less striking, change in the properties of C—H bonds, concomitant with the decreasing electropositive character of carbon in the polar bonds C—(hal.)
$\overset{+}{}\quad\overset{-}{}$
Moreover, the direction of this change continues unaltered through the hydrocarbon series on the right of Table 34. The observation that the shift in properties of the C—H bond between C$_2$H$_4$ and C$_2$H$_2$ is so much the largest in the whole series has prompted the suggestion that in C$_2$H$_2$ the polarity is C—H,
$\overset{-}{}\quad\overset{+}{}$
which is certainly in agreement with its acid function.

Table 34

	Force constant (C—H) (dyne/cm. × 10^{-5})	d(C—H) (A.)		Force constant (C—H) (dyne/cm. × 10^{-5})	d(C—H) (A.)
CH$_3$F	4·7	1·109	CH$_4$	5·0	1·094
CH$_3$Cl	4·9	1·101		(diff. 0·1)	(diff. 0·007)
CH$_3$Br	4·95	1·100	C$_2$H$_4$	5·1	1·087
CH$_3$I	5·0	1·100		(diff. 0·8)	(diff. 0·028)
			C$_2$H$_2$	5·9	1·059

It can hardly be doubted that the cause of the increase in electronegativity of carbon in the series CH$_4$, C$_2$H$_4$, C$_2$H$_2$ lies in the changing nature of the hybridization of the valency state. Between the hybrids [sp^2] and [sp] the degree of participation of the s orbit obviously increases much more than between [sp^3] and [sp^2]. Now it is very probable that the average distance from the nucleus of an electron in a carbon $2s$ orbit is less than such distance for the $2p$ orbit,* and it seems reasonable to infer that electrons in the hybrid $[sp]$ will be more firmly bound than those in $[sp^2]$ or $[sp^3]$.

Table 35. *Heats of combustion of cycloalkanes, and derived quantities†*

cyclo-Alkane	ΔH (corrected) (Cal.)	$\Delta H/n$ (Cal.)	Δ (Cal.)	$n\Delta$ (Cal.)
C_3H_6	499·5	166·5	+9·2	27·6
C_4H_8	655·6	163·9	+6·6	26·4
C_5H_{10}	793·0	158·6	+1·3	6·5
C_6H_{12}	943·8	157·3	—	—

† From Kaarsemaker and Coops, *Rec. trav. chim.* **71**, 261 (1952).

The cycloalkanes (cycloparaffins)

By the vigour of its chemical response to additives (e.g. H_2, Br_2, hydrogen halides, and H_2SO_4) the first member of this homologous series is not easily distinguished from the isomeric propylene, $CH_3.CH:CH_2$. In addition reactions a substituted c-propane ring breaks at a position such that the $1:3$ addition products obeys Markownikoff's rule. *cyclo*Butane is stable to Br_2 at normal temperature and can be hydrogenated to n-butane only at elevated temperature (e.g. Ni catalyst at 200°). Both c-propane and c-butane are more resistant to oxidants than the isomeric ethylenic hydrocarbons. From c-pentane onwards the members of the series behave as true *cyclo*paraffins, and are difficult to distinguish chemically from their analogues. Table 35 exhibits first the heats of combustion ΔH (corrected to perfect gas state), and in succeeding columns in order: the heats of combustion per CH_2 group ($\Delta H/n$); the differences Δ in $\Delta H/n$ from that for c-hexane; lastly a quantity $n\Delta$ which may be taken as a measure of the total 'strain' in the molecules, if it is assumed that there is no strain in c-hexane (see below).

To explain the structural data for c-propane and c-butane shown in Table 36 we shall need to expand the conceptions of hybridization of orbits in the C atom which have sufficed in the

* Moffitt, *Proc. Roy. Soc.* A, **202**, 534, 548 (1950). For detailed discussion of the C—H bond see Gent, *Quart. Rev.* II, 383 (1948).

preceding pages. If, from the s orbit and the three p orbits assumed in the valency state, we desire to form four *equivalent* hybrid orbits $[sp^3]$, that is orbits differing only in the spatial direction of their axes, then it is clear that the same proportion of the non-directional s orbit must be contributed to each hybrid. More precisely, in each of the expressions of the hybrid wave-functions, ψ_s must appear multiplied by the same fraction c_4. Similarly, in the three equivalent functions of $[sp^2]$ we have the common coefficient c_3, and in each of the two functions of $[sp]$, c_2: obviously $c_2 > c_3 > c_4$ (the explicit coefficients are respectively $1/\sqrt{2}$, $1/\sqrt{3}$, and $1/2$). There is, however, nothing in the theoretical basis of hybridization which compels us to construct sets of orbits which are all equivalent. For the purposes of argument we consider separately the two pairs of orbits into which the 'tetrahedral' orbits $[sp^3]$ may be arbitrarily divided. The planes defined by the two pairs of orbital axes are mutually perpendicular, and the symmetry of each pair of axes is C_{2v}: the inter-axial angles are, of course, 109°. If now equal amounts of s constituent are transferred from the members of pair (1) to those of pair (2), the members of each pair will remain equivalent to each other, but the transference must be compensated by a similar movement of p constituents in the opposite direction. Since in a process of the kind described the symmetry is pre-served, the two axial planes remain mutually perpendicular. The essential change produced is in the interaxial angles: that of pair (1) diminishes ($< 109°$) and that of pair (2) increases ($> 109°$). When the transfer process is carried to the limit and all s con-stituent is concentrated in pair (2), this pair becomes two equivalent $[sp]$ hybrids, with interaxial angle 180°, and pair (1) eventuates as two pure p orbits, with interaxial angle 90°. Hence no s, p hybrid orbits can be constructed with interaxial angles as low as 90°, and it follows that no such hybrid orbits can be directed along the internuclear lines in c-propane or c-butane.

We can only revert to the non-linear type of orbital overlap suggested by fig. 52 on p. 195 (and there rejected for *ethylene*). It remains to calculate, for the pairs of equivalent orbits on each C atom contributing mutually to the C, C links, what particular intermediate hybrid affords the maximum bond energy, which is mainly determined by the orbital overlap. From each of the three C atoms in c-propane two bonding orbitals, with interaxial angle θ, project in the plane of the ring. If the hybridization in the atoms were $[sp^3]$, $\theta = 109°$, and the angle of deviation between the orbital axes and the internuclear lines is $\delta = (109 - 60)/2 = 29·5°$. As the hybridization changes towards less s constituent (see above), θ and δ diminish, and the bonding energy increases until $\theta = 104°$, and $\delta = 22°$; after this point, as the hybrid orbits project less and less along their axes, overlap, and with it the energy of bonding, rapidly decreases, until when θ reaches its minimum at $90°$ ($\delta = 15°$), and the bonding orbits have pure p character, the overlap is small. With θ fixed at $104°$, the interaxial angle of the other two bonding orbits, which gives the angle HCH, automatically takes the value $116°$. For c-butane δ is only $9°$. The increase in C, C length consequent upon a bonding energy somewhat less than normal, when C, C is $1·54$A., is clearly shown in c-butane (Table 36); C, C in c-propane is less precisely known.

Table 36. *Structural properties of cycloalkanes*

cycloAlkane	C, C (A.)	C, H (A.)	∠CCC	∠CCH	∠HCH	Ref.
C_3H_6	—	—	60°	116·4°	118·2°	1
C_4H_8	1·568 ±0·02	1·098 ±0·04	90° (mean)	—	114±8°	2, 3
C_5H_{10}	1·54	1·09	109°	—	109·5°	1
C_6H_{12}	1·54	1·09	109°	109°	109°	1

(1) Hassel and Viervoll, *Acta Chem. Scand.* **1**, 149 (1947).
(2) Dunitz and Schomaker, *J. Chem. Physics*, **20**, 1703 (1952).
(3) Rathjens, Freeman, Gwinn and Pitzer, *J. Amer. Chem. Soc.* **75**, 5634 (1953).

The merit of the new theory over the formerly accepted 'strain theory' of Baeyer is that it helps to explain why the 'strain energy' is so small a fraction of the total energy: for c-propane it is but $3·3\%$ and for c-butane only $2·2\%$ of the total energy of formation from atoms (p. 142). The older theory contemplated that the tetrahedral C bonds were drastically bent until the interbond angle became 60° or 90° respectively. In the new model the strain energy arises from a slightly less favourable overlap than in the normal C, C bond, and from C—H bonds in which the s contribution is slightly larger than in [sp^3].*

Of the *cyclo*alkanes only the first two exist as truly planar structures. At normal temperature an out-of-plane vibration seems to operate in C_4H_8 (ref. 2, 3) but at low temperatures this would doubtless be quenched. In all the other members of the series the ring is 'puckered' in such a way as to preserve the tetrahedral valency angles and allow all bonds to be of the normal σ type. Puckering is too slight in *cyclo*pentane (plane polygon angle 108°) to have been detected, but in *cyclo*hexane it leads to two stereoisomeric molecules (the 'chair' and 'boat' forms). Beyond C_6H_{12} the puckering becomes increasingly complex and the possibility of numerous stereoisomers grows rapidly.

Benzenoid (aromatic) character

As an adaptation to modern ideas of an earlier suggestion by E. Bamberger† it has been proposed that the benzenoid properties of certain cyclic structures depend upon the formation of a special sextet of electrons.‡ Of the thirty valency electrons

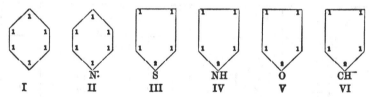

in benzene, C_6H_6 (I), twelve are allocated to the six single C—C bonds, another twelve to the six C—H bonds; the remaining six constitute the 'aromatic sextet', to which each carbon atom may be considered to contribute one electron. The position and

---

* For a full treatment of c-propane and c-butane see Coulson and Moffitt, *Phil. Mag.* **40**, 1 (1949).

† *Ber.* **24**, 1758 (1891); *Annalen*, **273**, 373 (1893).

‡ Armit and Robinson, *J. Chem. Soc.* **27**, 1604 (1925); Goss and Ingold, *ibid.* 1928, p. 1268.

structural function of the sextet is left undefined. It will be seen that two electrons are demanded from each of the 'hetero-atoms' in the 5-membered rings of thiophene, pyrrole and furane (III, IV and V). Now it is well known that thiophene, unlike its reduction product $\begin{array}{c} CH_2-CH_2 \\ | \\ CH_2-CH_2 \end{array}\rangle S$ and other organic compounds of sulphur, yields no sulphonium salts (p. 84), sulphoxide or sulphone; and furane, unlike pyrone and certain ethers, forms no oxonium salts (p. 80). Pyrrole is non-basic, while pyridine (II), the nitrogen of which is able to retain the essential bonding pair of electrons, is a relatively strong base. In *cyclo*pentadiene the demand for two electrons from the carbon of CH_2 can only be met by the drastic step of ionizing one of the C—H links in this group (VI). It is therefore not surprising that this hydrocarbon exhibits a predominantly olefinic (i.e. non-aromatic) character, but in readily yielding a potassium derivative $(C_5H_5)^-K^+$* it definitely displays the power of forming the sextet. It is significant that the open-chain di-olefine 1:4 pentadiene (Table 38, II) is not known to form metallic derivatives. Figs. I–VI have been primarily arranged in order of decreasing benzenoid character, which is seen to correspond with the order of increasing difficulty of extracting electrons from the hetero-atoms: extraction from CH_2, involving ionization, is the most difficult in the series.

Without losing sight of the obvious parallelism between the postulate of the aromatic sextet and the conclusion, drawn from physical theory (p. 197), of the presence in benzene of six 'mobile' electrons, we may with advantage interpret the sextet principle in terms of the theory of resonance, which was treated in an introductory way in Chapter IV (pp. 119 et seq.). We may here approach the subject from a different angle. Enough accurate data upon heats of formation, heats of combustion, bond energies, etc. have now been accumulated to make possible a calculation of heats of formation of a large number of chemical structures associated with definite bond-diagrams. Table 37 shows that large positive discrepancies ΔH may exist between the heats of

* Thiele, *Ber.* **34**, 68 (1901).

formation so calculated ($H_{calc.}$) and the observed heats ($H_{obs.}$). The structures for which $H_{calc.}$ has been evaluated are shown in the second column:

Table 37

$$\Delta H = H_{obs.} - H_{calc.}$$
(Cal.)

CO_2	$O{=}C{=}O$	38
CO_3^-	$O{=}C\begin{smallmatrix}O^-\\O^-\end{smallmatrix}$	44
Acids	$R_{alk.}{-}C\begin{smallmatrix}O\\OH\end{smallmatrix}$	15
	$R_{alk.}{-}C\begin{smallmatrix}O\\O_-\end{smallmatrix}$	>15
Esters	$R_{alk.}{-}C\begin{smallmatrix}O\\OR_{alk.}\end{smallmatrix}$	19
Urea	$O{=}C\begin{smallmatrix}NH_2\\NH_2\end{smallmatrix}$	33
Amides	$R_{alk.}{-}C\begin{smallmatrix}O\\NH_2\end{smallmatrix}$	20

For data on resonance energies see Wheland, '*Resonance in Organic Chemistry*', New York, 1955.

Although it is not possible in all cases to evaluate $H_{calc.}$ with precision, ΔH is far greater than the probable error in the calculation. Without exception structures showing this anomaly in the heat of formation, i.e. structures *more stable than the assigned bond-diagram would suggest,* can be represented by one or more alternative bond-diagrams or electron assignments. Such alternative diagrams may be distinct:

$$O{=}C{=}O \quad and \quad O^-{-}C{\equiv}O^+,$$

or *congruent*, i.e. superimposable after rotation of one form:

$$O^-{-}C{\equiv}O^+ \quad and \quad O^+{\equiv}C{-}O^-.$$

Examination of all the data shows clearly a regular and direct relation between the total number of bond-diagrams that can properly be constructed, and the value of ΔH, which reaches its largest values when congruent diagrams can be drawn. It may be noted that these facts and empirical relations would survive and invite explanation, even though the theory of resonance, which at present claims to explain them, were modified or discredited.

The theory of quantum mechanical resonance must not be regarded as merely an explanation *ad hoc* of such anomalies as are indicated above. On the contrary it claims to provide in principle a complete calculus of chemical union. The liberation of energy attendant on the reaction $A + B \to A - B$ is viewed as principally due to the possibility of writing more than one assignment of the two electrons that will ultimately participate in the bond:

(i) $A(1) + B(2)$ (ii) $A(2) + B(1)$

(1) and (2) designate the two electrons about to form the bond. Calculation of the liberated energy is effected (at least in principle) by considering the two systems (i) and (ii), with exchanged electrons and equal energies, as in sympathetic vibration or resonance. From this point of view covalent bond energy is essentially 'exchange' or resonance energy. It has, however, become the common practice to reserve the description 'resonance energy' for application to such cases of anomalous molecular stability as have been exemplified in Table 37 above. The additional energy of formation is considered to arise in these cases from resonance between the various aspects of the electron assignment. It is in harmony with classical no less than with modern principles that such resonance should be the stronger the more nearly equal the separate energies attributable to the units of resonating systems. As the energies of structures represented by congruent diagrams are necessarily precisely equal, it is in these cases that the resonance energy will be largest, as has already been remarked above. It has become customary to describe two or more distinguishable states of a system with equal or nearly equal energies as *degenerate* (see spectroscopic use of this term, p. 150). It is, however, necessary to exercise care in transferring the use of this term to the present subject, and particularly to recognize that congruent structures, although with

equal energies, are not necessarily in resonance. The diagrams
for the carboxylic group:

$$-C\overset{\displaystyle O}{\underset{\displaystyle OH}{\big\langle}} \qquad\qquad -C\overset{\displaystyle OH}{\underset{\displaystyle O}{\big\langle}}$$

are congruent but non-resonating, owing to the necessity of transferring the proton of the group from one oxygen atom to the
other. The actual resonating forms

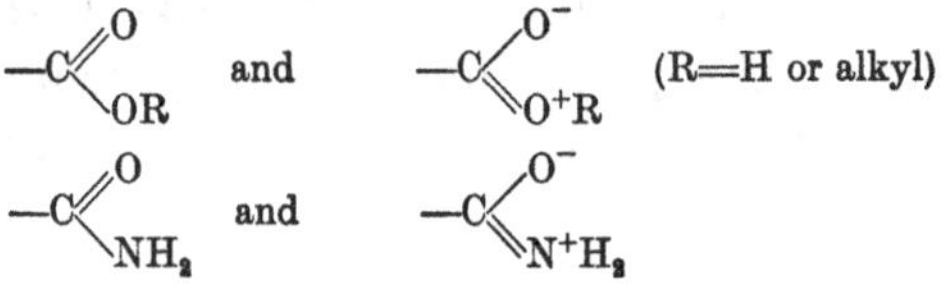

$$-C\overset{\displaystyle O}{\underset{\displaystyle OR}{\big\langle}} \quad \text{and} \quad -C\overset{\displaystyle O^-}{\underset{\displaystyle O^+R}{\big\langle}} \qquad \text{(R=H or alkyl)}$$

$$-C\overset{\displaystyle O}{\underset{\displaystyle NH_2}{\big\langle}} \quad \text{and} \quad -C\overset{\displaystyle O^-}{\underset{\displaystyle N^+H_2}{\big\langle}}$$

are non-congruent, and incompletely degenerate. For this reason
the two C, O distances in the carboxylic group are not equalized
(see oxalic acid, p. 199).* On the other hand, the carbonate ion
and the nitro group are examples of *complete* degeneracy, CO_3^-
being triply, and $-NO_2$ doubly, degenerate:

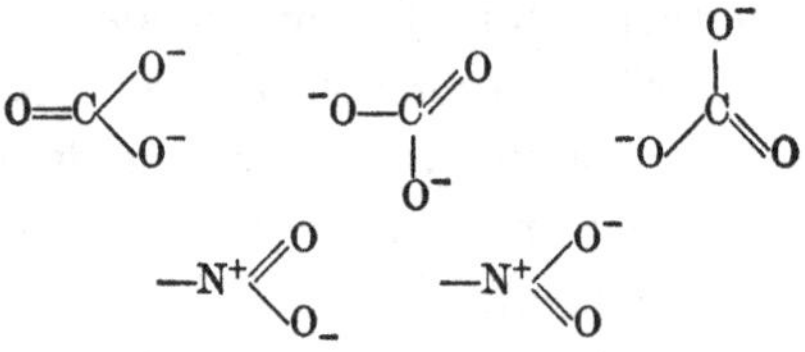

Energy arising from resonance between the two completely
degenerate 'Kekulé' structures is regarded as the principal
cause of the benzenoid stability of benzene and pyridine:

I   II

From the thermal data upon hydrogenation (p. 200) we see that
if in the reduction of benzene to dihydrobenzene a normal in-

---

* The acid *anion* is *completely* degenerate, with both C, O distances
equal, as in basic Be acetate, p. 56.

dependent ethylenic linkage were reduced, the energy change should be $-28\cdot6$ Cal., as in the change tetrahydrobenzene C_6H_{10} to *cyclo*hexane C_6H_{12}; in fact the reduction is endothermal, the energy change being $+7\cdot4$ Cal. (p. 200). Hence

$$\Delta H = H_{obs.} - H_{calc.} = 28\cdot6 + 7\cdot4 = 36 \text{ Cal.}$$

This energy is not wholly due to resonance between Kekulé forms I and II, but the contributions from resonance involving other possible bond-diagrams (excited forms) amount to less than 10 per cent of the whole.* The magnitude of ΔH for benzene may be compared with that of ΔH for other (open-chain) hydrocarbons, calculated also from Table 32:

Table 38

		ΔH (Cal.)
I	$CH_2{=}C{=}CH_2$	-11
II	$CH_2{=}CH.CH_2.CH{=}CH_2$	0
III	$CH_2{=}CH.CH{=}CH_2$	$3\cdot6$
IV	$CH_2{=}C\!\!-\!\!C{=}CH_2$ $\quad\;\;|\quad\;|$ $\quad\;\;CH_3\;\;CH_3$	$2\cdot8$
V	$HC{\Large\langle}^{CH-CH}_{CH_2-CH_2}{\Large\rangle}CH$	$1\cdot8$

It is clear that even for dihydrobenzene (V) the resonance energy is insignificant compared with that of benzene.

Figs. I–IV show the possible bond-diagrams for thiophene, in all of which the sulphur atom with its positive charge may be

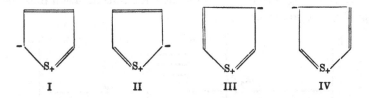

| I | II | III | IV |

said to exist in the 'sulphonium' condition. Analogous diagrams can be drawn for furane and pyrrole, in which respectively the

* Pauling and Wheland, *J. Chem. Physics*, **2**, 362 (1933).

oxygen takes the oxonium condition, and the nitrogen the ammonium condition. For the anion of *cyclo*pentadiene we have

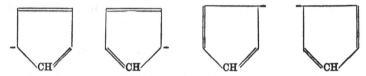

Values of ΔH for all the benzenoid structures considered above are collected in Table 39.

Table 39

	Form used for $H_{calc.}$	ΔH (Cal.)
Benzene		36
Pyridine		35
Thiophene		28
Pyrrole		22·5
Furane		21·5
c-Pentadiene		(Small)

In assessing the effect of resonance due consideration must be paid to the sequence:

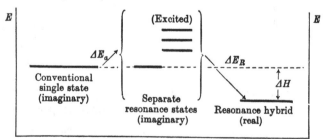

The importance of the part played by excited (or promoted) states in the resonance, and therefore the value of the excitation energy ΔE_a (per mol.), varies specifically. At the same time the excitation energy ΔE_a is self-limiting, because the more highly excited resonance states contribute little to the total resonance.

For benzene, excited states, e.g. ⟨structure⟩ are of minor concern,

$\Delta E_a \sim 0$ and ΔH is sensibly equal to the true resonance energy ΔE_R. On the other hand, for the 5-membered rings *all* the resonance states are excited states, and ΔH is *less* than the true resonance energy ΔE_R, particularly for c-pentadiene. The great difference in stability between the isoelectronic molecules N_2O and N_2CH_2 (diazomethane) may well be due to the considerable excitation energy of the resonance structures, such as $N\equiv\overset{+}{N}-\overset{-}{C}H_2$, which are required for the latter molecule (see p. 130). Allene is exceptional in its heat of formation, which is less by 11 Cal. (Table 38, I) than that calculated for its conventional structure $CH_2{=}C{=}CH_2$, assumed to contain ordinary ethylenic links. Although allene is isoelectronic with CO_2, resonance units corresponding with those for the oxide (e.g. $O^-{-}C\equiv O^+$) cannot be formed without the invalid assumption of 5-covalent carbon. The stereochemistry of allene derivatives,* and the fact that the $C{=}C$ distances (1·33) and force-constants ($9·8 \times 10^5$) are practically identical for ethylene, allene and propylene,† exclude bond-diagrams other than that above. The cause of the lessened stability lies very probably in the low degree of orbital hybridization in the central carbon atom. Instead of $[sp^2]\,p$ as in ethylene (see p. 195) we require the less stable $[sp]\,p^2$. The classical stereochemistry is ensured by the axes of the two pure p-orbits being mutually at right angles; $[sp]$ gives two collinear bonds (as in $BeCl_2$, pp. 56 and 115) at right angles to the p-bonds.

---

* Mills and Maitland, *J. Chem. Soc.* 1936, p. 987.

† Pauling and Brockway, *J. Amer. Chem. Soc.* 59, 1223 (1937); Eyster, *J. Chem. Physics,* 6, 580 (1938); Thompson and Linnett, *J. Chem. Soc.* 1937, p. 1384.

In regard to the homologues of benzene it would be expected that the introduction of substituents, themselves without potential resonating forms, e.g. alkyl groups, would detract from the degeneracy of the parent hydrocarbon by modifying its symmetry, and so decrease the resonance energy. Such an effect would be in harmony with the greater general chemical reactivity of the homologues of benzene compared with that of benzene itself. The effect of such substituents is, however, in general to *increase* ΔH, as may be judged by the heats of complete hydrogenation in Table 40:

Table 40

	Heat of complete hydrogenation (Cal.)	Change in ΔH
Benzene	$49\cdot8\pm0\cdot15$	—
Ethylbenzene	$48\cdot9\pm0\cdot10$	$0\cdot9$
o-Xylene	$47\cdot2\pm0\cdot20$	$2\cdot6$
Mesitylene	$47\cdot6\pm0\cdot20$	$2\cdot2$

(From Kistiakowski, *loc. cit.* p. 200.)

The almost complete agreement between the calculated and observed heats of formation of the paraffin hydrocarbons lends no support to the assumption that appreciable energy can arise from resonance forms of alkyl groups, although it must be admitted that data on heats of combustion are hardly sufficiently precise to give final disproof of this possibility. From what has been said above the explanation may lie in a relation between ΔE_a and ΔE_R for these homologues, different from that in benzene. Penney, in an illuminating discussion of the aromatic hydrocarbons,* implicitly assumes (by adopting the same 'bond order' for all benzene homologues) that the resonance of benzene undergoes no modification by the entry of alkyl substituents. It may be noted that the entry of methyl groups into butadiene (Table 38, III and IV) does reduce ΔH and presumably also the resonance energy.

Interesting examples of the relations of side-chains to nuclear resonance are provided by certain heterocyclic ring structures

* Penney, 1937, *loc. cit.* p. 208.

that have been fully examined by X-ray methods. In cyanuric triazide (I), with 'Kekulé' formula, all the constituent atoms lie in one plane, and two C, N distances are found in the ring − 1·38 and 1·31, corresponding to C—N and C═N respectively in formula I: further, C, (N_3) is also 1·38.*

The inequality of the ring distances, while the 'single-bond' distance (1·38) is less than the standard length of C—N (1·47— Table 5, p. 44), indicates *incomplete* resonance in the nucleus. The diminished value of the side-chain 'single-bond' C—(N_3) shows that the incomplete nuclear resonance is part of a more general resonance in which the azide group —N_3 participates. In melamine (II), on the contrary, as in benzene, all the bond-lengths in the planar rings are equal: C, C in benzene = 1·39; C, N in melamine = 1·35.†

Cyanuric tri-amide, like benzene, therefore displays *complete* resonance between complementary Kekulé formulae. We have in this phenomenon a means of determining the otherwise difficultly accessible bond-length of C═N, for, assuming that C, C in benzene bears the same relation to C—C (1·54) and C═C (1·33) as C, N in melamine (II) does to C—N (1·47) and C═N (x), we have

$$\frac{1·47 - 1·35}{1·47 - x} = \frac{1·54 - 1·39}{1·54 - 1·33},$$

whence $x = 1·30$.

Even in melamine (II) the C, N distance in the side-chain is less than that of C—N, and we must presume that the —NH_2 groups take some part in a general resonance, in the form

$$H_2N^+ \!\!=\!\! C\!\!<$$

---

* Knaggs, *Proc. Roy. Soc.* A **150**, 576 (1935).

† Hughes, *J. Amer. Chem. Soc.* **63**, 1737 (1941).

Inorganic compounds exhibiting complete 'Kekulé' resonance are triborine-tri-imine (III),* and phosphonitrile chloride (IV)†

Returning now to the principle of the aromatic sextet, we see that its ease of formation provides a guide to the extent to which resonance can stabilize a cyclic structure. In polynuclear aromatic compounds, such as naphthalene, quinoline, anthracene and acridine (I–IV), it is clear that the maximum contributions must fall short of providing sextets for all the rings, the

contributions being 5 per ring for bi-nuclear and 4 per ring for tri-nuclear structures. Such limitations are strikingly reflected in the chemistry of polynuclear compounds. Unlike benzene all the above compounds can be reduced in well-defined stages, in which the power of the reducing process must be increased.

Although in tetrahydro-naphthalene (III) the conjoined hydro-aromatic ring might be expected to impose a slight restraint on

---

* Bauer, *J. Amer. Chem. Soc.* **60**, 524 (1938).

† Brockway and Wright, *ibid.* **65**, 1551 (1943).

the 'Kekulé' resonance of the other ring, this ring is nevertheless reduced catalytically (to decalin, IV) in one indivisible stage like benzene. Intermediate stages of hydrogenation may, however, be distinguished by the use of HI as reducing agent. In dihydro-anthracene (VI) the restraint on the two other rings is probably greater and even catalytic hydrogenation proceeds by numerous stages:

V ⟶ VI $\xrightarrow{\text{catalytic reduction}}$ tetrahydride

↓ do.

'perhydride' $C_{14}H_{24}$ ⟵ do. decahydride ⟵ do. octahydride

The chemistry of dimethyl-γ-pyrone (VII) provides an interesting illustration of the interplay between benzenoid and other conjugated structures:

VII $+ H^+ + X^- \rightleftharpoons$ VIII $+ X^-$

In the oxonium cation (VIII) of dimethyl pyrone the ready ionization of the H atom from the OH group occasions considerable reversion, even in acid solution, to the form VII; consequently this pyrone functions as a very weak base ($k_b = 3\cdot0 \times 10^{-14}$; cf. urea, $k_b = 1\cdot5 \times 10^{-14}$), in spite of the resonance of the benzenoid cation.* On the contrary the CH_3 of the methoxy group of the methylated pyrone (IX) cannot appreciably ionize, and the whole of the pyrone retains the benzenoid form, with $k_b = $ ca. 10^{-5}; cf. NH_3, $k_b = 1\cdot5 \times 10^{-5}$:

IX

When substituents are present which in themselves offer alternative 'resonance' structures, e.g. $-NO_2$, $-CHO$, $-COOH$,

* Cf. Gibbs, Johnson and Hughes, *J. Amer. Chem. Soc.* **52**, 4895 (1930).

—Cl, —NH$_2$, —OH, the composition of the resonance group of states will be profoundly modified, and it is probable that excited states will play a more prominent part in the hybridization. From a detailed exploration of such altered conditions it should be possible to elucidate the well-known empirical laws of aromatic substitution.

The substituents —NH$_2$, —OH and —Cl may be expected to give rise to the following triplets of excited states:

Contracted symbol

(NH$_2$)$^+$	(NH$_2$)$^+$	(NH$_2$)$^+$	(NH$_2$)$^+$ (−)/(−)(−)
(OH)$^+$	(OH)$^+$	(OH)$^+$	(OH)$^+$ (−)/(−)(−)
(Cl)$^+$	(Cl)$^+$	(Cl)$^+$	(Cl)$^+$ (−)/(−)(−)

In the case of aniline and phenol the conversion of the nitrogen and oxygen atoms to the ammonium and oxonium condition need cause no surprise; it will be noticed that the bond-diagrams for these compounds closely resemble those for pyrrole and furane (p. 211). The assumption of doubly-bonded, positively charged halogen is less easily accepted, but receives strong support from the fact that $C_{aryl.}$, Cl is shorter than $C_{alk.}$, Cl (as in CH$_3$Cl), and that a similar shortening is found in aliphatic compounds when Cl or Br is bound to 'ethylenic' carbon, as in vinyl chloride Cl.CH : CH$_2$ and bromostyrene C$_6$H$_5$.CH : CHBr.* The resistance to hydrolysis of halogenated benzene and of the other types of compound mentioned is also in harmony with the assumption. From the point of view of the theory of molecular orbits the change from single to double bond merely means the eviction of electrons in anti-bonding orbits in the original link C—Cl to orbits in the link C—C.† It is recognized that 'direct'

---

* *Ann. Reports*, 1937, p. 206; 1940, p. 76.

† Cf. Price and Tutte on the ultra-violet spectrum of CHCl:CHCl (*Proc. Roy. Soc.* A, **174**, 218 (1940)).

substituents in the benzene nucleus are electron-seeking or electrophilic, e.g. —NO_2 or —SO_3H. It is therefore evident that the groups —NH_2, —OH and —Cl will exert an ortho-para directing influence of strength proportional to the importance of the above structures in the resonance group, and in turn to the ease with which the extra-nuclear double bond essential to the structures can be supposed to be formed. Such a deduction is in complete agreement with the observed order for power of ortho-para direction, viz. —NH_2 > —OH ≫ —Cl. The substituent exerting apparently the greatest power of ortho-para direction is —O^-, from the phenoxide C_6H_5ONa. This is also to be expected, since a prior negative charge on the oxygen atom obviously facilitates greatly the change:

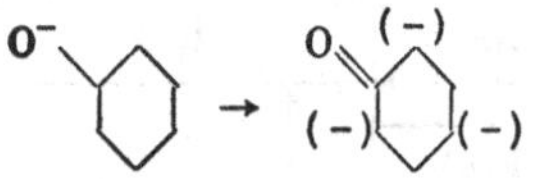

Well-known reactions such as the easy tri-halogenation of aniline and phenol in the 2 : 4 : 6 positions also fall into line.

As an example of the directing effect of another type of substituent we may consider the aldehyde group —CHO:

Conventional structure

Contracted symbol

(1) (2) (3)

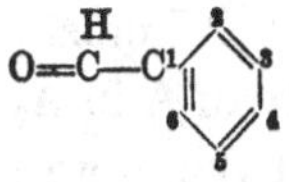

For benzaldehyde we may postulate the excited structures 1–3, the effect of which upon the nucleus is indicated in the contracted form. Electrophilic substituents will now be prevented from taking the ortho-para positions, but may adopt the meta-positions; these latter sites are not, however, as in the former case of ortho-para substitution, *specially* favoured by the presence of considerable negative charge. This deduction is again entirely in

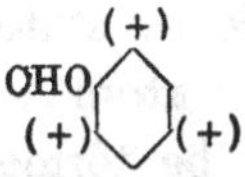

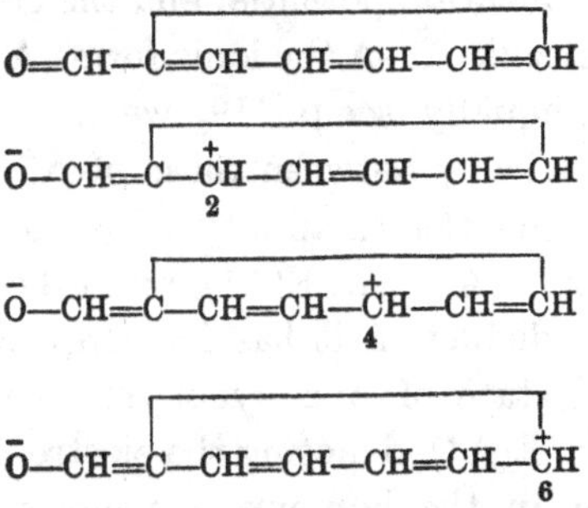

accordance with experience, in that meta-substitution is in general
a much more sluggish process than ortho-para substitution, and
gives an impression of 'pis aller'. We may thus perceive the general
outlines of a theory of aromatic substitution, but space will not
permit this complicated subject to be further pursued here.

SECTION III. THE ELECTRO-AFFINITY (ELECTRO-NEGATIVITY) OF COMBINED ATOMS

If for the univalent atom A we take as reference state the anion
A^-, the electro-affinity (E.A.) of which we set arbitrarily at zero,
then the affinities of the free atom and of the cation A^+ can be
displayed on an energy scale thus:

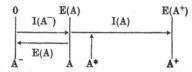

When, as is more customary, the free atom is assumed as
reference state, the separation of A^- from A^+ on the scale is seen
to be $I(A) + E(A)$, where $I(A)$ and $E(A)$ are respectively the
(first) ionization potential and the electro-affinity of A. Since to
the molecule A—A the ionic forms $A^-.A^+$ and $A^+.A^-$ must contribute equally (see p. 119), we infer that the state of A in A_2 lies
symmetrically between those of A^+ and A^- and therefore that
$E(A^*)$, the electro-affinity of A combined in A_2, lies midway
between $E(A^-)$ and $E(A^+)$, and takes the value $[I(A) + E(A)]/2$.
In this deduction it has for simplicity been assumed that the
normal state of A is synonymous with its valency state (see
below). $E(A^*)$, denoting the electro-affinity of the atom A when
present in the homonuclear molecule A_2 may be termed its
standard E.A. (electro-negativity) for the combined state.

Clearly the standard E.A. varies specifically for different
(univalent) atoms. In the heteronuclear molecule A—B the
inequality of $E(A)^*$ and $E(B)^*$ is resolved by a decrease of the
higher, say $E(A^*)$, towards that of the state A^-, and a corresponding increase of $E(B^*)$ towards the state B^+ (see scale
above). In this way the *effective* electro-affinities are restored to

balance. The extent of the necessary shifts, and therefore the importance of the ionic form $A^-.B^+$ in the molecule AB, is proportional to the difference of standard E.A.

Serious obstacles to an easy evaluation of standard E.A. of multivalent atoms arise from at least two causes: (1) the valency states for which I and E are required are not usually physically observable, (2) the effect of the hybridization of orbits in the valency states must be allowed for. Even in treating univalent atoms some difficulties arise. In an exact calculation of the energy of formation of H_2 excited states of the atoms had to be included as correction terms, and thus must play some part in the valency state of hydrogen; the halogen cations X^+ must be given, for the purpose of assigning the correct I, not the ground state $s \uparrow\downarrow .p_x \uparrow\downarrow .p_y \uparrow p_z \uparrow$, ^3P, but a mixture of excited states, such as $s \uparrow\downarrow .p_x \uparrow\downarrow .p_y \uparrow\downarrow$, ^1D and $s \uparrow\downarrow .p_x \uparrow\downarrow .p_z \uparrow\downarrow$, ^1S. Mulliken* has attempted to surmount all these difficulties, and the results of his calculations are summarized in Table 41 (1).

Table 41. *Electro-affinities of combined atoms*
(in electron-volts)

(The values are applicable only to *single* bonds.)

	H	Li	Be $[sp]$	B $[sp^2]$	C $[sp^3]$	N	O	F
(1)	7·1	2·9	4·7	6·2	8·2	8·5	10·0	12·7
(2)	7·1	—	—	—	8·5	9·5	10·6	12·1
		Na	K	Rb	Cs	I	Br	Cl
(1)		2·6	2·2	2·0	1·9	8·2	9·1	9·8
(2)		—	—	—	—	8·1	9·0	9·4

Other scales, founded more empirically, have been devised. The best known of these† derives from the assumption that the energy of the *hypothetical* covalent bond A—B (in which both ionic forms are given *equal* weight) is the mean of the bond energies of A—A and B—B. The excess Δ_{AB} of the observed bond energy over this estimate is attributed to covalent-ionic

* *J. Chem. Physics*, **2**, 782 (1934).
† Pauling, *J. Amer. Chem. Soc.* **54**, 3570 (1932).

resonance, and taken to be proportional to the quantity $[E(A^*)—E(B^*)]^2$. The results of this method, adjusted so that $E(H^*)$ is common to both scales, are shown in Table 40 (2).

Standard E.A. certainly decreases with increasing atomic number in groups I and VII of the periodic system, and there is no reason to doubt that a similar decrease will exist in the intervening multivalent groups. It will be noticed that when the C atom adopts the state $[sp^3]$ a small polarity $C_-—H_+$ is predicted for C,H bonds, while other theoretical and experimental evidence indicates a contrary resultant polarity.* Such apparently conflicting results do not necessarily invalidate the data of Table 41, but rather indicate that covalent-ionic resonance (or unequal sharing of bonding electrons) is not the sole cause of the existence of molecular dipole moment. Other discernible causes, already suggested on pp. 141 and 187, may, when the E.A. are not dissimilar, markedly affect the strength of the moment and even reverse its direction, as indeed appears to be the case in C,H bonds.†

SECTION IV. HYDROGEN BONDS

Numerous examples are known in which we are compelled to presume that two of the elements N, O or F are linked by the agency of 'hydrogen'. Compounds dependent on such bonds occur not only in the crystalline state but also in the liquid and gaseous states. Before the rigour of Pauli's exclusion principle (p. 147) was fully appreciated it seemed plausible to assume that hydrogen could co-ordinate two groups while the elements of Series II co-ordinated four groups. When the impossibility of expanding the capacity of the 1s-orbits of hydrogen beyond two electrons was fully recognized, the nature of the bond had to be re-examined.

The simplest and clearest examples of a hydrogen bond are found in the hydrofluoride ion HF_2^-, and the gaseous polymers $(HF)_n$. The relevant facts have been recorded on pp. 88, 89. In

---

* Gent, *Quart. Rev.* II, 383 (1948).
† Coulson, *Trans. Faraday Soc.* **38**, 433 (1942).

the monofluorides such as KF, etc. the radius of F^- is established as 1·33, while the radius of F— (covalent radius) is 0·59–0·62 (p. 45). The F,F distances in HF_2^- and in $(HF)_n$, although variable, all lie between twice these radii. There appears to be only one possible explanation, viz. a symmetrical ionic complex of the form $F^-.H^+.F^-$.* One would expect that, if the complex (FHF)$^-$ occurring in crystals of hydrofluorides were purely ionic, the F,F distances, being themselves regulated merely by *Coulomb* forces, would be sensitive to the nature and particularly the radii of the associated cations. The data on p. 88 suggest that the cationic 'stretching' power may be in the order of decreasing cationic radius, Na$\gg$(NH$_4$,K), as expected. The F,F distances (2·5–2·7) in the gaseous and solid $(HF)_n$ appear to be too large, but these are the least well-authenticated of the measurements.

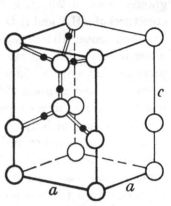

Fig. 55. The hexagonal unit cell of ice. O, O; (●, H): O=Zn=S in fig. 6. a=4·53A., c=7·41A. For NH$_4$F, a=4·39, c=7·02.

Both NH$_4$F† and ice‡ exhibit the wurtzite structure (fig. 6, p. 41); the hexagonal unit cell of ice is shown in fig. 55. All the other NH$_4$ halides show the NaCl or CsCl structure (fig. 11, p. 43). The volume of the hexagonal unit cell of NH$_4$F, which contains two molecules, is

$$\sqrt{3}ca^2/2 = (\sqrt{3} \times (4\cdot39)^2 \times 7\cdot02)/2 = 118 \text{ cu. A.}$$

Hence the molecular volume is 59 cu. A. In the dense packed ionic lattices of KF and RbF (NaCl structure) the volumes of the cubical unit cells each containing four molecules are respectively

* Peterson and Levy, *J. Chem. Physics*, **20**, 704 (1952); Richards, *Quart. Rev.* x, 480 (1956).

† Zachariasen, *Z. phys. Chem.* **127**, 218 (1927); Kronberg and Harker, *J. Chem. Physics*, **10**, 309 (1942). (Hydrazine fluoride N$_2$H$_4$.(HF)$_2$ is also discussed in the second communication.)

‡ W. H. Bragg, *Proc. Phys. Soc.* **34**, 98 (1922); Barnes, *Proc. Roy. Soc.* A, **125**, 670 (1929).

$(5\cdot32)^3 = 150$ cu. A. and $(5\cdot64)^3 = 179$ cu. A. As the radius of NH_4^+ ($1\cdot43$) lies between that of K ($1\cdot33$) and that of Rb ($1\cdot49$) we should expect that a dense packed lattice of NH_4^+ and F^- ions would result in a volume of about 40 cu. A. per molecule of NH_4F. A calculation upon similar lines shows that if ice were a dense packed lattice of H_2O molecules it would have a specific gravity of about $2\cdot0$. It is therefore evident that these crystal structures of NH_4F and H_2O are very 'open', and may be likened to a (3-dimensional) lace pattern. As the specific gravity of liquid water is only a little greater than that of ice ($0\cdot916$) we must presume that only a slightly denser packing results even on melting. It would appear to follow that strong and tetrahedrally directed bonding forces must be present in both crystals. The sum of the radii of NH_4^+ and F^- is $1\cdot43 + 1\cdot33 = 2\cdot76$. The N, F distance found in the crystal of NH_4F is $2\cdot61$, the contraction of $0\cdot15$ being also evidence of bonding rather than mere physical contact. In ice the O, O distance is $2\cdot76$, and the open structure of ice described above is *prima facie* evidence for this being the length of an OH, O bond, shown in fig. 55. In 'heavy' ice at $-50°$ C. O, D is $1\cdot01$ A.,* and is thus only slightly longer than O, D in the free molecule D_2O (O, D $= 0\cdot96$ A.). This result was obtained by neutron diffraction, in which deuterium compounds afford much clearer indications than those of 'light' hydrogen.

The closest approach of two oxygen atoms bonded in *separate* molecules or ions may be estimated in several ways. The structures of certain simple silicates and phosphates, such as olivine (Mg_2SiO_4), triphylite ($Li(Fe,Mn)PO_4$), and the pure phosphate Li_3PO_4, can be regarded as essentially close-packed assemblies of O atoms bonded in tetrahedra, with Si or P set in the tetrahedral spaces, and the cations in octahedral spaces. The practical equality of the (orthorhombic) cell dimensions in these examples strengthens this hypothesis:

	a	b	c	
Mg_2SiO_4	$4\cdot75$	$10\cdot20$	$5\cdot99$	Gossner and Strunz, *Z. Krist.* **83**,
$Li(Fe, Mn)PO_4$	$4\cdot67$	$10\cdot34$	$6\cdot00$	415 (1932).
Li_3PO_4	$4\cdot86$	$10\cdot26$	$6\cdot07$	Laves and Zambonini, *Z. Krist.* **83**, 26 (1932).

* Peterson and Levy, *Acta Cryst.* **10**, 70 (1957).

In such compounds the minimal O, O distance between two SiO_4 or PO_4 tetrahedra is 2·8, giving an 'external' radius of 1·4 for bonded oxygen.

All the Mg and Mn salts listed below* have the cubic NaCl structure, and it is seen that as the radius of the anion increases from $O^=$ to $Se^=$ the sides (a) of the unit cells become equalized. This must mean that in the selenides the anions are in contact (see fig. 56) and $r(Se^=)$ is $5·45/2\sqrt{2} = 1·93$ A.

MgO	a (A.) = 4·20		MnO	a (A.) = 4·43
MgS	a (A.) = 5·19		MnS	a (A.) = 5·21
MgSe	a (A.) = 5·45		MnSe	a (A.) = 5·45

K_2O a = 6·44: K_2S a = 7·39: K_2Se a = 7·68.

The potassium salts† also have a cubical cell but the anti-fluorite structure (p. 43). Hence the difference $r(Se^=) - r(O^=)$ is $[a(K_2O) - a(K_2Se)]\sqrt{3/4} = 1·24 \times \sqrt{3/4} = 0·54$. The radius $r(O^=)$ is therefore 1·39 A. A close agreement is to be expected between this estimate and that given previously for the 'external' radius of bonded oxygen, since the latter is electronically in a state very similar to that of $O^=$.

It may hence be considered that the closest approach of two O atoms in a crystal lattice corresponds to an interionic distance of not less than 2·8 A.

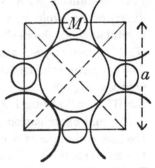

Fig. 56. A face of the cubic cells of MgSe and MnSe.

The distance of closest approach of OH^- ions or OH groups is not easy to establish. In fact the easiest course here is to examine examples where the structure seems to compel us to recognize bonding between two OH groups. In the layer lattices of the alkaline earth hydroxides $Mg(OH)_2$ and $Ca(OH)_2$, and in the similar lattice of LiOH, the OH ions adopt an approximately close-packed arrangement, i.e. no two OH ions lie one directly above the other. Such a 'staggered' arrangement for free OH^-

* Wyckoff, *Crystal Structures*, vol. I, Table III, 2 (3).
† Wyckoff, *Crystal Structures*, vol. I, Table IV, 2.

ions is to be expected in view of their strongly polar nature along the OH axis. On the contrary, both in FeO(OH), lepidocrocite, and in $Al(OH)_3$, hydrargillite, the OH groups are found 'end-on' in pairs, a mutual disposition highly improbable if the paired groups were not in some way constrained to take up this position. The O, O distances in these hydroxides are found to be as follows:

	O, O		O, O
LiOH	3·60	FeO(OH)	2·70
$Mg(OH)_2$	3·22	$Al(OH)_3$	2·78
$Ca(OH)_2$	3·36		

(From Bernal and Megaw, *Proc. Roy. Soc.* A, **151**, 384 (1935).)

Thus the end-on position goes not with the increased distances that the repulsions of free OH groups would demand, but with greatly reduced distances. To presume some form of bonding seems inescapable.

It will be recalled that in the phosphate Li_3PO_4 (and the silicate Mg_2SiO_4) the XO_4 ions approach so that the shortest O, O distance in two tetrahedra is 2·8. Now if the H ion in KH_2PO_4 played an ordinary cationic role there is no reason to expect a less distance, but crystal analysis shows that the distance is shortened to 2·50: in $NaHCO_3$ two oxygen atoms in every CO_3^- ion are each distant 2·55 from others in neighbouring CO_3^- ions. In $(NH_4)_2H_3IO_6$ every O in the IO_6 ion is held 2·60 from another in a second IO_6 anion, the H^+/O ratio being 1:2 as in the acid phosphate.

A recent examination of KH_2PO_4 (at normal temperature) by the new technique of neutron diffraction* has shown that in the O—H $\cdots$ O bonds linking the PO_4 tetrahedra together the H atoms lie not more than 0·18 A. from the central position, and hence that the minimum length for O—H is 1·07 A. It is possible that in $NaHCO_3$ (fig. 59) and H_2SeO_3 (fig. 60) the 'short' hydrogen bonds also contain a very extended O—H link. Formerly it was sought to explain the short bond in oxalic acid dihydrate by formulating the compound as di-oxonium oxalate,

$$(H_3O^+)_2.C_2O_4^-,$$

but this supposition has been discredited.†

* Bacon and Pease, *Proc. Roy. Soc.* A, **220**, 397 (1953).

† Richards and Smith, *Trans. Faraday Soc.* **47**, 1261 (1951); *ibid.* **48**, 675 (1952).

Table 42. Hydrogen bonds

Inorganic acids, and acid salts:			*Carboxylic acids:*		
Ref.	XO···HOX	O, O (A.)	Ref.	>CO···HO.CO<	O, O (A.)
1	H_3BO_3	2·71	12	HCO_2H*	2·73
2	H_2SeO_3	2·56, 2·60	12	CH_3CO_2H*	2·76
3	HIO_3	2·78	12	CF_3CO_2H*	2·76
4	KH_2PO_4	2·50	13	$H_2C_2O_4$	2·71
5	KH_2AsO_4	2·54	14	$C_2H_4(CO_2H)_2$	2·64
6	$NaHCO_3$	2·55		*Dimeric forms	
6 (a)	$Na_3H(CO_3)_2.2H_2O$	2·53			
7	$(NH_4)_2H_3IO_6$	2·60			

Hydrates:			*Hydroxy-compounds:*		
	XO···H_2O			OH···OH	
			15	Ice	2·76
			16	H_2O_2	2·78
8	$Na_2CO_3.H_2O$	2·70	17	Resorcinol	
8 (a)	$Na_3H(CO_3)_2.2H_2O$	2·75		$(m\text{-}C_6H_4(OH)_2)$	2·70
9	$H_2C_2O_4.2H_2O$	2·52	18	Pentaerythritol	
				$(C(CH_2OH)_4)$	2·69

M.H_2O···H_2O and			*Nitrogen compounds:*		
XO···H_2O				NH_3^+···OC<	N, O (A.)
10	$CuSO_4.5H_2O$	2·75 (mean)	19	Glycine	
11	$NiSO_4.7H_2O$	2·72 (mean)		$(NH_2CH_2CO_2H)$	2·76, 2·88
				OH···OC<	O, O(A.)
			20	Hyperol	
				$(\text{Urea-}H_2O_2)$	2·63

1 Zachariasen, *Z. Krist.* 88, 150 (1934).
2 Wells and Bailey, *J. Chem. Soc.* 1949, p. 1282.
3 Helmholz and Rogers, *J. Amer. Chem. Soc.* 63, 278 (1941).
4 West, *Z. Krist.* 74, 306 (1930).
5 Helmholz and Levine, *J. Amer. Chem. Soc.* 64, 354 (1942).
6 Zachariasen, *J. Chem. Physics*, 1, 634 (1933).
6 (a) Brown, Peiser and Turner-Jones, *Acta Cryst.* 2, 167 (1949).
7 Helmholz, *J. Amer. Chem. Soc.* 59, 2036 (1937).
8 Harper, *Z. Krist.* 95, 266 (1936).
8 (a) See 6 (a).
9 Dunitz and Robertson, *J. Chem. Soc.* 1947, p. 142.
10 Beevers and Lipson, *Proc. Roy. Soc.* A, 146, 570 (1934).
11 Beevers and Lipson, *Z. Krist.* 83, 123 (1932).
12 Karle and Brockway, *J. Amer. Chem. Soc.* 66, 574 (1944).
13 Cox, Dougill and Jeffery, *J. Chem. Soc.* 1952, p. 4854.
14 Morrison and Robertson, *J. Chem. Soc.* 1949, p. 987.
15 Barnes, *Proc. Roy. Soc.* A, 125, 670 (1929).
16 Abrahams, Collin and Lipscomb, *Acta Cryst.* 4, 15 (1950).
17 Robertson, *Proc. Roy. Soc.* A, 157, 79 (1936). A, 167, 122 and 136 (1938).
18 Llewellyn, Cox and Goodwin, *J. Chem. Soc.* 1937, p. 883.
19 Albrecht and Corey, *J. Amer. Chem. Soc.* 61, 1087 (1939).
20 Lu, Hughes and Giguère, *J. Amer. Chem. Soc.* 63, 1507 (1941).

It will be clear from Table 42 that in the majority of known hydrogen bonds the O, O distance is about 2·75A. As stated on

p. 224 the O, H distances of the hydrogen bonds in solid D_2O, in which O, O is 2·76A. are respectively O'D, 1·01 and O''D, 1·75A. The bonding hydrogen is therefore very unsymmetrically placed, and remains almost as close to its 'own' oxygen as in a free D_2O molecule (O, D = 0·96A.). Although such direct evidence has not yet been obtained for 'long' hydrogen bonds in other hydroxylic compounds, the small shift of the characteristic OH vibration frequency in their infra-red absorption spectra*

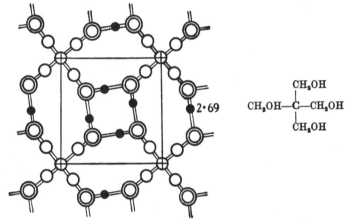

Fig. 57. The structure of pentaerythritol, $C(CH_2OH)_4$.
$\oplus$, C of $C(CH_2OH)_4$; $\circledcirc$, O; $\bullet$, H of OH.

affords clear evidence for an unsymmetrical arrangement like that in ice. If we assume that the orbits of the oxygen atom in a water molecule are hybridized in the tetrahedral type [sp^3], then two doubly-occupied hybrid orbits will project in directions, separated by 109°, away from the OH bonds, and in a plane perpendicular to that of these bonds (cf. p. 141). In the atom there will therefore be two localities of negative charge so placed that if they attract hydrogen atoms of other water molecules the observed tetrahedral structure of ice must result. On melting, the regular open structure of ice becomes somewhat disorganized, and a slightly closer packing becomes possible, corresponding to an increase of density from 0·916 to 1·00 g. At 4° for

* Sutherland, *Trans. Faraday Soc.* **36**, 889 (1940).

H_2O, and 11° for D_2O thermal expansion begins to overtake this effect, but even at the normal boiling-point intermolecular bonding still persists and accounts for the large latent heat of evaporation. There can be no doubt that similar bonding is present in liquid hydroxy compounds such as the alcohols, but to a lesser degree than in water owing to the reduction of the H/O ratio to unity.

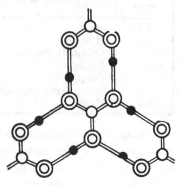

Glycine (amino-acetic acid) forms a layer structure of parallel chains, of which the units are the 'zwitter-ions' $NH_3^+CH_2CO_2^-$. Each —NH_3^+ group forms two hydrogen bonds $NH_3^+...CO_2^-$, one, of length 2·75 A., to an adjacent ion in its chain, and a second of

Fig. 58. The structure of boric acid $B(OH)_3$. $\bigcirc$, B; $\circledcirc$, O; ●, H.

length 2·88 A. which cross-links the chains. A similar general constitution may be presumed for protein aggregates.

The possibility of the formation of hydrogen bonds often

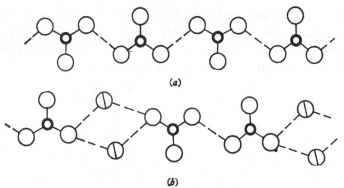

(a)

(b)

Fig. 59. Infinite chain anions in (a) $NaHCO_3$, (b) $NaHCO_3.Na_2CO_3.2H_2O$. ●, C; $\bigcirc$, O; $\oplus$, O in H_2O; - - -, H bond.

determines crystal structure, as shown in fig. 58 for H_3BO_3 and in fig. 57 for pentaerythritol. The type of structure adopted depends primarily on the number of hydrogen bonds per molecule that can be formed, and this in turn is determined by the appropriate H/O ratio in the compound. Table 43 summarizes these considerations.

Table 43

H/O ratio	Examples	H-bond/O ratio	Structure type
2:1	Ice	4:1	Wurtzite
1:1	H_3BO_3 (fig. 58) $C(CH_2OH)_4$ (fig. 57)	2:1	Layer
1:2	KH_2PO_4 $(NH_4)_2H_3IO_6$	1:1	3-Dimensional network
1:3	HIO_3, $NaHCO_3$ (fig. 59)	2:3	Infinite chains
2:3	H_2SeO_3 (fig. 60)	4:3	Double layer

Among organic compounds we may distinguish three different consequences of the formation of hydrogen bonds: (a) intermolecular bonding resulting in 'unlimited association', as in the alcohols and phenols; (b) intermolecular bonding resulting in dimerization. This type is characteristic of monocarboxylic acids in non-polar solvents, or as vapours. The simple units are linked in pairs, thus:

$$R—C\underset{O—H—O}{\overset{O—H—O}{\big<}}\big>C—R$$

(in *crystalline* carboxylic acids hydrogen bonding leads to a macromolecular structure); (c) intramolecular bonding: a characteristic OH absorption band at 7100 cm.$^{-1}$ is suppressed in salicylaldehyde (I) and o-nitrophenol (II):*

The stereochemical positions in the o-isomers allow a bond of length 2·7 to be formed as shown. In the m- and p-isomers of both substances the OH frequency is strongly marked. The greater volatility of the enol form of acetoacetic ester compared

* Hilbert, Wulf, Hendricks and Liddel, *Nature*, 135, 147 (1935).

with that of the keto form is unexpected, but may be attributed to an O, OH bond:

$$CH_3-C \underset{\underset{O \cdots H \cdots O}{|}}{\overset{\overset{CH}{\diagup}\diagdown}{}} C.OEt$$

Unlike mixtures of most organic liquids those of chloroform with diethyl ether or acetone show strong *negative* deviations from Raoult's law and at certain compositions yield azeotropic mixtures of maximum boiling-point. It is generally accepted that the great difference in strength between acetic acid and its

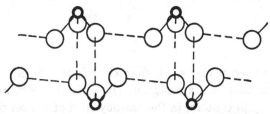

Fig. 60. Double-layer structure in selenious acid, H_2SeO_3.
o, Se; O, O; - - -, H bond.

trichloro derivative is occasioned by the drift of electrons across the molecule in response to the attraction of the strongly electronegative halogen atoms: as a consequence the carboxyl hydrogen becomes more positive and more readily ionized. It is plausible that for a similar reason the hydrogen atom in chloroform is sufficiently positive to form hydrogen bonds with the oxygen of ether or acetone; the union resulting would explain the deviation from Raoult's law.*

Compared with most bond energies that of the 'long' hydrogen bond is low: estimates lie between 5 and 10 Cal. for O—H—O and N—H—O. The energy of the ionic bond in HF_2^- is much higher, at about 30 Cal., and it is possible that the energy of the 'short' hydrogen bond (e.g. in acid salts) is of intermediate magnitude.

* Huggins, Pimentel and Shoolery, *J. Chem. Physics*, **23**, 1244 (1955).

SECTION V. ELECTRON-DEFICIENT SYSTEMS

We shall define electron deficiency as the condition occurring when the total number of electrons in the valency shells of the n atoms in a molecule or macromolecular system is less than the number $2(n-1)$. Under this condition there will be insufficient electrons to form pair-bonds all of which are localized between two atoms. It is necessary to include all the valency shell electrons and not only the (unpaired) electrons defining the primary valency. Thus Al_2Cl_6, with the constitution

$$\underset{Cl}{\overset{Cl}{>}}Al\underset{\underset{+}{Cl}}{\overset{\overset{+}{Cl}}{<}}Al\underset{Cl}{\overset{Cl}{<}}$$

is not regarded as electron-deficient, while the dimeric $Al_2(CH_3)_6$ does come into such a category. The molecules BF_3 and $AlCl_3$, although both contain, in the valency shell of the central atom, a vacant orbit readily occupied by appropriate co-ordination (p. 59), are not electron deficient in the special sense implied above.

The solid alkali metals provide probably the most obvious example of electron deficiency. The N electrons in the valency shells of N alkali-metal atoms would suffice to form only $N/2$ discrete molecules M_2, if localized electron pair bonds were taken as the only means of chemical union. Such diatomic molecules are in fact rather unstable (see Table 29, p. 179) and occur to only a very limited extent in the vapours of the alkali metals. An accepted theory of the constitution of these metals implies the opposite extreme of delocalization, by assuming that every electron binds all the atoms and is associated with a wave-function co-extensive with the metal system. Examples of electron-deficient molecules are provided at present mainly by hydrogen compounds of Groups III elements: e.g. all the boron hydrides (p. 60); digallane, Ga_2H_6; aluminium methyl, $Al_2(CH_3)_6$; aluminium hydride, $(AlH_3)_n$; and the borohydrides $Ga(CH_3)_2BH_4$, $Be(BH_4)_2$, $Al(BH_4)_3$ (p. 60). We select the simplest boron hydride B_2H_6 as a type for detailed discussion.

Undoubtedly the most signal advance in our knowledge of diborane during the past decade has been the final establishment of its structure. Formerly, on the basis of insufficiently refined electron diffraction data, its structure was supposed to be similar to that of ethane (1)

$$\begin{array}{ccc}
\text{H} & & \text{H}\\
\text{H—B} & \text{——} & \text{B—H}\\
\text{H} & & \text{H}
\end{array} \qquad
\begin{array}{ccc}
\text{H} & \text{H} & \text{H}\\
& \text{B} \quad \text{B} &\\
\text{H} & \text{H} & \text{H}
\end{array}$$

(1) (2)

but such a formulation was found to be incompatible with its infra-red and Raman spectra, the study of which first suggested the symmetrical 'bridged' structure (2), now finally confirmed by the technique of nuclear magnetic resonance.* The most recently obtained electron diffraction evidence† yields the following dimensions for model (2): $B, B = 1\cdot770 \pm 0\cdot013$ A.; B, H (terminal) $= 1\cdot187 \pm 0\cdot030$ A. (cf. B, H in $BH_3CO = 1\cdot20$ A.); B, H (bridge) $= 1\cdot334 \pm 0\cdot027$ A.; $\angle HBH$ (terminal) $= 121\cdot5 \pm 7\cdot5°$ $\angle HBH$ (central) $= ca.\ 97°$. It is clear that some form of 'hydrogen' bond is concerned in diborane, but it cannot have the electrostatic character of O—H—O or N—H—O already discussed, since the electro-affinities of H and B are too similar (Table 41). The former view that bonds of the type O—H$\cdots$O could be explained by resonance between the forms OH—O and O—HO was discredited by the discovery of the unsymmetry of the H atom in the 'long' hydrogen bond, since for resonance to be operative only the electron configuration may be varied, around the nuclei assumed fixed in position. The data given above show that a similar obstacle must hinder the interpretation of diborane as a resonance between pairs of BH_3 units

for B, H in the resonance forms is presumably $ca.\ 1\cdot19$ A. while in the actual molecule B_2H_6, B, H in the 'bridge' is $ca.\ 1\cdot33$ A.

* Shoolery, *Discuss. Faraday Soc.* 19, 215 (1955).
† Hedberg and Schomaker, *J. Amer. Chem. Soc.* 73, 1482 (1951); see also Bauer, *Chem. Rev.* 31, 46 (1942).

The most recent propositions start from the assumption that in a unit $>BH_2$ with normal B, H bonds, there would remain one singly-occupied orbit and one unoccupied orbit, of an approximately $[sp^3]$ hybridization (see p. 204). A pair of these orbits, one from each B atom and both directed to one side of the B—B axis, form with the $1s$ of a hydrogen atom equidistant from each boron atom a total of three molecular 3-centred orbits, of which only the lowest is bonding (cf. fig. 45). An exactly similar 3-centred bonding orbit can be constructed symmetrically on the

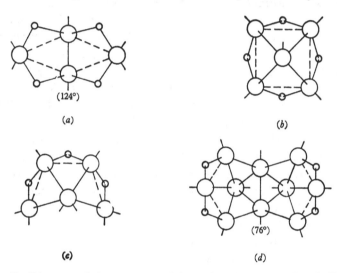

Fig. 61. Diagrams of the structures of boron hydrides. (a) B_4H_{10}, (b) B_5H_9, (c) B_5H_{11}, (d) $B_{10}H_{14}$. O, BH; O, BH_2; o, H (bridge). (Broken lines indicate the planes in which the boron atoms lie.)

opposite side of the B, B axis, and the total of four electrons, two from 2B and two from 2H, complement these bonding orbits.

With the present exception of B_6H_{10} the structures of all the boranes listed on p. 60 have been successfully explored by diffraction methods, and it is noteworthy that the relatively low scattering power of boron has permitted the positions of hydrogen to be deduced with unusual certainty. In all the hydrides hydrogen is bound in two ways: (i) in normal groups, BH_2 or BH, with B, H ≤ 1·25 A., and (ii) in the angular bridge, B.··^{H}··.B, with B, H $ca.$ 1·35 A. 'Constitutional' formulae to record the groups present may be written as below:

B_2H_6 $(BH_2)_2.H_2$

B_4H_{10} $(BH_2)_2.(BH)_2.H_4$

B_5H_9 $[(BH)_4.(H_4)].BH$

B_5H_{11} $[(BH_2)_2.(BH)_2.H_3].BH_2$

$B_{10}H_{14}$ $[(BH)_8.(H_4)].(BH)_2$

The arrangements of the groups in the molecules are indicated in the diagrams of fig. 61. In B_5H_9 the B atoms in four BH groups lie in a square with H bridges underlying its sides: the fifth B atom, also in a BH group, is placed vertically above the centre of the square and apparently binds equally all the B atoms below it. The other pentaborane, B_5H_{11}, is derivable from B_5H_9 by opening one side of the square, when the bent chain so formed ($\angle BBB = 112°$) requires terminal BH_2 groups and only three bridges: there is again a fifth B atom, in a BH_2 group, bound to the four atoms beneath. In decaborane, $B_{10}H_{14}$, which like B_4H_{10} is a folded molecule, the ten B atoms are arranged as two regular pentagonal pyramids, joined at an edge and with the apices directed outwards; an angle of 76° separates the basal planes of the pyramids. The two B atoms placed, in BH groups, at the pyramidal apices exert a 'common' binding action over *five* other B atoms. The binding of the boron atoms in these hydrides evidently poses further problems in electron deficiency, to which multi-centred orbits again appear to provide the only solution (Longuet-Higgins, 1957, *loc. cit. infra*).

It is a striking fact that for all the boranes B_nH_m the sum $3n + m$ is even, so that like the hydrocarbons they all contain an even number of electrons, which must all be paired in spin, since all boranes are diamagnetic. Since the instability of BH_3 could be connected with the tendency of a boron atom to attain full quadri-covalency (as in B_2H_6), it is surprising that neither $B(CH_3)_3$ nor BCl_3 shows any tendency to become dimerized like the corresponding Al compound, in which Al attains quadri-covalency. In the greater number of its complex compounds, however, Al is seen in 6-co-ordination, and it is therefore not improbable that its covalent borohydride (p. 61) should be represented as

$$\text{Al}\left[\begin{array}{c}\text{H}\cdots\\ \quad\quad\text{B}\end{array}\begin{array}{c}\text{H}\\ \text{H}\end{array}\right]_3$$

and its solid hydride $(\text{AlH}_3)_n$ as a layer structure of hexagonal graphitic pattern wherein each C, C bond is replaced by $\text{Al}(\text{H}_2)\text{Al}$.

(References: Bauer, *Chem. Rev.* **31**, 43 (1942); Longuet-Higgins and Bell, *J. Chem. Soc.* 1943, p. 250; Longuet-Higgins, *J. Chem. Soc.* 1946, p. 139; *Quart. Rev.* XI, 121 (1957); Stone, *Quart. Rev.* IX, 174 (1955).)

APPENDIX

I. CONSTANTS, SIGNS, CONVENTIONS, ETC.

Fundamental physical constants *

F = Faraday's constant = $96,489 \pm 7$ coulombs.

c = velocity of light = $2 \cdot 99776 \times 10^{10}$ cm. sec.$^{-1}$

e = electronic charge = $4 \cdot 8025 \pm 0 \cdot 001 \times 10^{-10}$ e.s.u. (from wave-length of X-radiation).

$N = \dfrac{F}{e} \times \dfrac{c}{10} = 6 \cdot 023 \times 10^{23}$ molecules per g.-mol.

R = constant of the gas equation

$$pV = RT = 1 \cdot 986 \text{ cal./mol./degree.}$$

$\dfrac{h}{e}$ = $1 \cdot 3793 \pm 0 \cdot 0002 \times 10^{-17}$ e.s.u.; c.g.s. (from continuous X-ray spectrum).

h = Planck's constant = $\dfrac{h}{e} \times e = 6 \cdot 624 \pm 0 \cdot 002 \times 10^{-27}$ erg. sec.

Physical symbols

A.   Angstrom unit of length = 1×10^{-8} cm.

Z Atomic number.

z Ionic charge (valency of ion).

λ Wave-length (cm., or Angstrom units).

ν Frequency of radiation (sec.$^{-1}$).

ν' Wave-number ($1/\lambda$, cm.$^{-1}$).

R Rydberg's constant (spectroscopy).

$|\,X\,|$ Modulus of X, or absolute value of X, irrespective of its sign.

Energy: units and conventions

Cal.: 1 kilo-calorie = 1000 calories.

Electron-volt: 1 ev. is the energy acquired by an electron in passing freely through 1 volt potential difference. If the whole energy so gained is transferred to a molecule B (or an atom A) the energy received per g.-mol. of B (or g.-atom of A) is 23.05 Cal., i.e.

   1 ev. per molecule (atom) = $23 \cdot 05$ Cal. per g.-mol. (g.-atom).

Reciprocal cm.: The use of this unit is commonly confined to the evaluation of energy levels in an atom or molecule. To a

* Birge, *Rev. Mod. phys.* **13**, 233 (1941).

difference $\Delta\nu$ of frequency, or $\Delta\nu'$ of wave-number, there corresponds the difference of energy

$$\Delta E = h\Delta\nu = hc\Delta\nu';$$

and the difference $\Delta\nu' \equiv 2\cdot857$ (cal./g.-mol.) $\times \Delta\nu'$.

$$1 \text{ ev.} = 8067 \text{ cm.}^{-1}.$$

$\pm\Delta H$ Endothermal $(+)$ or exothermal $(-)$ heat change.

$\pm\Delta H_f$ Heat of formation (from common forms of elements).

$\pm\Delta F$ Change of free energy (absorption *positive*, liberation *negative*).

Chemical signs and symbols

(A), AB   Valency of A in the compound AB.

A, B   Distance (in A. units) between the nuclei of A and B in a molecule containing both atoms.

$\angle$ABC   The angle subtended at the nucleus of B by a line joining the nuclei of A and C, in a molecule containing all three elements.

A^+—B^- (or $A \rightarrow B$)   Co-ionic bond (A as 'donor').

A_+—B_- Resultant polarity of linkage.

II. *Bond-lengths*

(given or referred to in the text, but not collected in Tables 5, 10, 24 or 29)

		PAGE			PAGE
Ag—N	2·04	*55*	P—P	2·21	*74*
B—H	1·20	*233*	P$_2$	1·895	*74*
B—F	1·30	*136*	P—F	1·57	*75*
B, O (BO$_3^-$)	1·36	*136*	P—Cl	2·04	*153*
Be—O	1·65	*81*	P, O (P$_4$O$_6$)	1·62	*74*
C$_2$	1·312	*132*	P, O (P$_4$O$_{10}$)	$\begin{cases}1{\cdot}39\\1{\cdot}62\end{cases}$	*74*
F—O	1·38	*88*			
Cl—O (Cl$_2$O)	1·70	*89*	P=O (POX$_3$)	1·54–1·58	*75*
Cl, O (ClO$_3^-$)	1·48	*91*	P=S (PSX$_3$)	1·85–1·94	*75*
I—O (IO$_2$F$_2^-$)	1·93	*95*	S—S	2·08	*42*
I, O (IO$_4^-$)	1·79	*97*	S$_2$	1·88	*74*
I—F (IO$_2$F$_2^-$)	2·00	*95*	S—F	1·58	*85*
I—Cl	2·38	*95*	S—Cl	1·99	*83, 84*
H—H	0·741	*45*	S=O (SO$_2$, SO$_3$)	1·45	*84, 85*
H—Cl	1·275	*46*	S, O (SO$_3^-$)	1·39	*85, 91*
O$_3$	1·28	*128*			

For bonds between C, N, O or H see Tables 5 and 10: for XO$_4$ ions Table 24.

SUBJECT AND FORMULA INDEX